Agricultural and Food Waste Management - Innovative Solutions and Sustainable Practices

Edited by Anca Corina Fărcaş

Published in London, United Kingdom

Agricultural and Food Waste Management - Innovative Solutions and Sustainable Practices
http://dx.doi.org/10.5772/intechopen.1005991
Edited by Anca Corina Fărcaş

Contributors
Annu Khatri, Awais Munir, Defe Rameck, Hilaas Ahmad Peerzada, Iftikhar Ahmed, Indu Shekhar Thakur, Jay Shankar Singh, Krishan Kumar, Maria Ameen, Matutu Danmore, Mohamed Soliman Elshikh, Mudasir Ali, Muhammad Iqbal, Muhammad Saqlain Zaheer, Muhstaq Ahmad, Mutote Karren, Prateek Singh, Shabir Ahmad

Notice

Statements and opinions expressed in the chapters are these of the individual contributors and not necessarily those of the editors or publisher. No responsibility is accepted for the accuracy of information contained in the published chapters. The publisher assumes no responsibility for any damage or injury to persons or property arising out of the use of any materials, instructions, methods or ideas contained in the book.

First published in London, United Kingdom, 2025 by IntechOpen
IntechOpen is the global imprint of INTECHOPEN LIMITED, registered in England and Wales, registration number: 11086078, 167-169 Great Portland Street, London, W1W 5PF, United Kingdom

For EU product safety concerns: IN TECH d.o.o., Prolaz Marije Krucifikse Kozulić 3, 51000 Rijeka, Croatia, info@intechopen.com or visit our website at intechopen.com.

British Library Cataloguing-in-Publication Data
A catalogue record for this book is available from the British Library

Agricultural and Food Waste Management - Innovative Solutions and Sustainable Practices
Edited by Anca Corina Fărcaş
p. cm.
Print ISBN 978-1-83635-043-9
Online ISBN 978-1-83635-042-2
eBook (PDF) ISBN 978-1-83635-044-6

If disposing of this product, please recycle the paper responsibly.

Meet the editor

Anca Corina Fărcaş holds a Ph.D. in Biotechnology and is currently an associate professor and researcher in the Department of Food Science at the University of Agricultural Sciences and Veterinary Medicine Cluj-Napoca, Romania. Her research is centered on the sustainable valorization of agro-industrial by-products, with a strong focus on the recovery of bioactive compounds and the development of eco-innovative food ingredients and materials. She has led and contributed to numerous projects addressing circular economy strategies in the agri-food sector, aiming to reduce waste, enhance resource efficiency, and support low-impact food processing technologies. Her interdisciplinary work bridges biochemistry, food safety, and environmental responsibility, with a particular emphasis on integrating scientific research into practical, scalable solutions for managing agricultural and food waste. Through both academic and applied initiatives, she promotes sustainable practices that align with current global goals for food security, climate resilience, and the zero-waste concept.

Contents

Preface

Agricultural and food waste management has become a defining issue for the future of sustainable food systems. With the intensification of global agriculture and food production, vast quantities of biomass, ranging from crop residues and animal manure to processing by-products and unsold food, are discarded annually, often without consideration for their latent value. In this context, waste is no longer a by-product to be eliminated but a strategic resource to be recovered, transformed, and reintegrated into circular value chains.

This volume, *Agricultural and Food Waste Management – Innovative Solutions and Sustainable Practices*, brings together timely insights from multidisciplinary perspectives on how agricultural waste can be managed more sustainably and efficiently valorized. The chapters explore a variety of themes, including resource recovery through composting, biochar production, and bioenergy generation; policy frameworks supporting circular practices; and the role of innovation and local community engagement in closing nutrient and economic loops.

In line with the global drive toward low-carbon economies, this collection highlights the relevance of bio-based alternatives and renewable energy derived from agricultural residues. Emerging technologies, such as bioconversion, microbial treatment, and smart systems for waste monitoring, are also discussed as promising tools for optimizing waste utilization. Beyond technical aspects, the book addresses broader implications for food security, ecosystem services, and sustainable rural development.

The environmental consequences of unsustainable practices, such as crop residue burning, are also addressed, offering evidence-based discussions on their impact on air quality, soil health, and urban resilience. Equally important is the focus on rural development models that utilize agricultural waste as a lever for socio-economic progress, highlighting that sustainability must be both ecological and equitable.

Throughout the volume, the authors emphasize that addressing agricultural waste requires integrated solutions that are rooted in both scientific knowledge and practical implementation. By advancing novel approaches, this collection reinforces the role of circular economy principles in reshaping the agri-food landscape, minimizing losses, optimizing resource utilization, and contributing to global food security and climate change mitigation goals.

It is hoped that this book will serve as a valuable reference for researchers, professionals, and decision-makers committed to fostering a more resilient, sustainable, and innovative agri-food system.

Anca C. Fărcaş
Faculty of Food Science and Technology,
University of Agricultural Sciences and Veterinary Medicine,
Cluj-Napoca, Romania

Chapter 1

Agricultural Waste Management for Food Security and Sustainability

Krishan Kumar, Annu Khatri and Indu Shekhar Thakur

Abstract

Agricultural waste management is a crucial aspect of sustainable agriculture, directly influencing food security, environmental health, and economic viability. Agricultural activities generate diverse waste streams, including crop residues, animal manure, food processing by-products, and aquaculture waste. If not managed properly, these wastes contribute to soil and water pollution, greenhouse gas emissions, biodiversity loss, and economic inefficiencies. However, innovative waste management strategies, such as composting, vermicomposting, biogas production, and biochar application, can transform agricultural waste into valuable resources. Technological advancements, including artificial intelligence (AI), the Internet of Things (IoT), and biotechnology, play a vital role in optimizing waste recycling and valorization processes. Additionally, integrating circular economy principles into agricultural systems can enhance resource efficiency, reduce dependency on chemical fertilizers, and promote sustainable food production. Effective policy frameworks, international guidelines, and active participation from governments, NGOs, and farmers are necessary for the large-scale adoption of sustainable waste management practices. Addressing key challenges, such as the lack of awareness, financial constraints, and infrastructural limitations, will be essential for transitioning toward a zero-waste agricultural system that supports both environmental sustainability and global food security.

Keywords: agricultural waste, food security, sustainability, waste valorization, circular economy

1. Introduction

Agricultural waste management refers to the systematic handling, treatment, and utilization of waste generated from agricultural activities to minimize environmental degradation and maximize resource efficiency. Agricultural waste includes a wide range of materials such as crop residues, animal manure, food processing by-products, and aquaculture waste. Improper disposal of these waste streams can lead to severe environmental consequences, including soil degradation, water contamination, greenhouse gas (GHG) emissions, and biodiversity loss. Additionally, inefficient waste management contributes to economic inefficiencies by increasing dependency on chemical fertilizers, elevating production costs, and limiting the potential of waste-derived value-added products. To address these concerns, agricultural waste management focuses on converting waste into useful resources. Technologies such as composting, vermicomposting, anaerobic digestion (biogas production), and biochar application have been widely recognized for their ability to enhance soil fertility, improve energy recovery, and mitigate pollution. In recent years, advancements in artificial intelligence (AI), the Internet of Things (IoT), and biotechnology have further optimized waste monitoring, recycling, and valorization processes [1]. By integrating these innovative technologies, agricultural systems can operate more efficiently, reducing waste accumulation and environmental impacts. The scope of agricultural waste management extends beyond environmental conservation to include economic and social sustainability. By implementing circular economy principles, agricultural waste can be transformed into organic fertilizers, biofuels, and other sustainable alternatives that reduce reliance on synthetic inputs. This shift not only strengthens farm productivity but also contributes to climate change mitigation by lowering carbon emissions associated with agricultural activities. Furthermore, sustainable waste management practices align with global sustainability objectives such as the United Nations Sustainable Development Goals (SDGs), particularly SDG 2 (Zero Hunger), SDG 12 (Responsible Consumption and Production), and SDG 13 (Climate Action). Addressing agricultural waste effectively requires interdisciplinary approaches, technological innovations, and supportive policy frameworks to ensure a transition toward a more sustainable and zero-waste agricultural system. Food security, as defined by the Food and Agriculture Organization (FAO), relies on the availability, accessibility, utilization, and stability of food supplies. The mismanagement of agricultural waste poses a direct threat to food security by depleting soil nutrients, contaminating water sources, and reducing the productivity of agricultural land. As modern agricultural systems become increasingly dependent on external inputs such as chemical fertilizers and synthetic pesticides, the improper disposal of organic waste exacerbates soil degradation and weakens the resilience of food production systems [2].

Effective agricultural waste management, however, ensures nutrient recycling, soil health preservation, and long-term agricultural productivity, ultimately supporting food security at local and global scales. Beyond food production, sustainable agricultural waste management plays a pivotal role in climate resilience and environmental sustainability. Agricultural waste, particularly livestock manure, crop residues, and food processing by-products, contributes significantly to methane (CH_4) and nitrous oxide (N_2O) emissions, two potent greenhouse gases responsible for global warming. Strategies such as biogas production, biochar application, and composting can mitigate these emissions while improving soil carbon sequestration. Additionally, proper waste management prevents the leaching of harmful substances into water bodies, reducing risks of water pollution and eutrophication, which can have devastating impacts on aquatic ecosystems and food supply chains. The economic implications of agricultural waste management are also significant. By converting agricultural residues into valuable products such as organic fertilizers, biofuels, and livestock feed, farmers can reduce input costs, enhance profitability, and contribute to sustainable rural economies [3].

This is particularly critical for small-scale farmers in developing regions, where limited access to chemical fertilizers and energy sources can hinder agricultural productivity. Furthermore, global initiatives such as the Paris Agreement and the European Green Deal emphasize the importance of sustainable waste management in reducing carbon footprints and promoting regenerative agricultural practices. Governments, international organizations, and private-sector stakeholders are increasingly recognizing the need for financial incentives, regulatory policies, and technological support to enhance agricultural waste recycling and circular economy adoption. By integrating efficient waste management strategies into agricultural systems, societies can achieve multiple sustainability goals, including improved food security, economic resilience, and climate adaptation. The development of innovative waste valorization approaches, combined with supportive policies and education initiatives, is essential for fostering a sustainable agricultural system that minimizes waste while maximizing environmental and economic benefits [4]. This study aims to evaluate sustainable agricultural waste management strategies and their impact on food security, environmental sustainability, and economic viability. Effective waste management can enhance soil fertility, reduce pollution, and mitigate greenhouse gas emissions, thereby contributing to climate resilience and sustainable agriculture. By examining techniques such as composting, biogas production, and biochar application, this study explores their potential for nutrient recycling and waste valorization. Additionally, emerging technologies like artificial intelligence (AI), the Internet of Things (IoT), and biotechnology are assessed for their role in optimizing waste management processes. Furthermore, this study investigates the economic benefits of waste-to-resource approaches, particularly their potential to reduce farming costs and improve rural livelihoods. Policy frameworks, financial incentives, and regulatory measures are also analyzed to understand their effectiveness in promoting large-scale waste management adoption. Ultimately, this research seeks to identify challenges and propose strategies for transitioning toward a zero-waste agricultural system that supports global sustainability efforts.

2. Types and classification of agricultural waste

Agricultural waste includes crop residues, animal manure, and food processing and aquaculture waste, each requiring specific management strategies. Crop residues, such as straw and husks, can be used for composting, biochar production, and bioenergy instead of open-field burning. Animal manure, rich in nutrients, is utilized through composting, biogas production, and organic fertilizer applications. Food processing and aquaculture waste can be repurposed for bioenergy, composting, or extracting valuable bioactive compounds [5].

2.1 Crop residues and their management

Crop residues are the plant materials left behind after harvesting, including straws, husks, stalks, leaves, and roots. These residues represent a significant fraction of agricultural waste and, if not properly managed, can contribute to environmental pollution and soil degradation. Open-field burning, a common disposal practice, releases particulate matter, carbon dioxide (CO_2), methane (CH_4), and nitrogen oxides (NO_x), contributing to air pollution and climate change. Therefore, adopting sustainable management strategies is crucial for enhancing soil fertility, reducing emissions, and promoting circular agricultural systems. Various types of crop residues (wheat straw, rice husks, and corn stalks) and the environmental issues caused by improper management, such as open-field burning and soil degradation, are shown in **Figure 1** [6]. Incorporating crop residues into the soil is one of the most effective methods of enhancing soil health. Residue retention improves soil organic matter, increases microbial activity, and enhances water retention capacity, thereby reducing erosion and improving long-term agricultural productivity. Additionally, crop residues can be used as mulch to suppress weeds, regulate soil temperature, and conserve moisture. Biochar production through pyrolysis offers another sustainable approach. Biochar, a stable carbon-rich material, improves soil structure, enhances nutrient availability, and sequesters carbon, reducing atmospheric CO_2 levels. Furthermore, anaerobic digestion of crop residues can generate biogas, providing a renewable energy source while reducing organic waste accumulation. Residues such as rice husks and wheat straw can also be repurposed for bioenergy production through direct combustion or gasification. Moreover, they serve as raw materials for livestock feed, mushroom cultivation, and industrial applications, including paper and bio-composite production. Effective crop residue management requires the integration of sustainable practices with technological advancements, such as precision agriculture and microbial decomposition accelerators. Government policies and incentives promoting residue recycling and alternative uses play a crucial role in minimizing open-field burning and encouraging farmers to adopt environmentally friendly practices. Sustainable crop residue utilization is essential for improving soil health, reducing greenhouse gas emissions, and promoting resource-efficient agricultural systems [7].

2.2 Animal manure: Composition and utilization

Animal manure is a by-product of livestock farming, composed primarily of feces, urine, bedding materials (such as straw or sawdust), and residual feed. The key

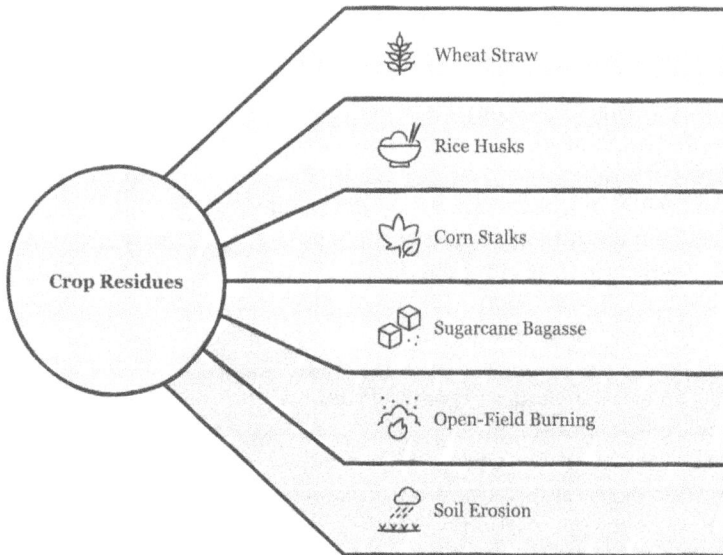

Figure 1.
Classification of crop residues and the environmental impacts of improper management, including open-field burning and soil degradation.

components of animal manure, such as feces, urine, bedding materials, and leftover feed shown in **Figure 2**. It is a rich source of organic matter, nitrogen (N), phosphorus (P), potassium (K), and micronutrients essential for plant growth. The composition of manure varies depending on animal species, diet, and management practices, with cattle manure generally containing higher nitrogen content, while poultry manure is rich in phosphorus. Properly managing and utilizing animal manure is crucial for maintaining soil health and reducing environmental impacts. When applied to agricultural land, manure serves as an organic fertilizer, improving soil structure, enhancing water retention, and promoting microbial activity. The nutrients in manure contribute to soil fertility, reducing the need for synthetic fertilizers. However, improper manure management can lead to nutrient imbalances, groundwater contamination from nutrient leaching, and the emission of methane (CH_4) and nitrous oxide (N_2O), both potent greenhouse gases [8]. Various manure management strategies have been developed to optimize its utilization. Composting is one of the most common methods, where manure is aerated and broken down by microorganisms to reduce pathogens and odor, transforming it into a stable, nutrient-rich organic fertilizer. Anaerobic digestion, in which manure is decomposed in the absence of oxygen, produces biogas (methane), which can be used for energy generation, while also producing a nutrient-rich digestate for use as fertilizer. The use of manure as a soil amendment requires careful management to prevent over-application, which can lead to nutrient runoff and environmental pollution. Integrated manure management systems, combined with technological innovations like precision nutrient application, can help optimize manure's benefits, improving crop yields while minimizing negative environmental impacts. Through these sustainable practices, animal manure contributes to the circular economy, reducing reliance on synthetic fertilizers and promoting sustainable farming [9].

2.3 Food processing and aquaculture waste

Food processing and aquaculture industries generate substantial organic waste, which, if not managed effectively, contributes to environmental pollution and resource inefficiencies. Food processing waste includes fruit and vegetable peels, seed husks, cereal bran, dairy by-products, and meat processing residues. Aquaculture waste consists of uneaten feed, fish excreta, processing discards (fish scales, bones, and viscera), and sludge from aquaculture systems. These waste streams contain valuable organic matter, nutrients, and bioactive compounds that can be repurposed

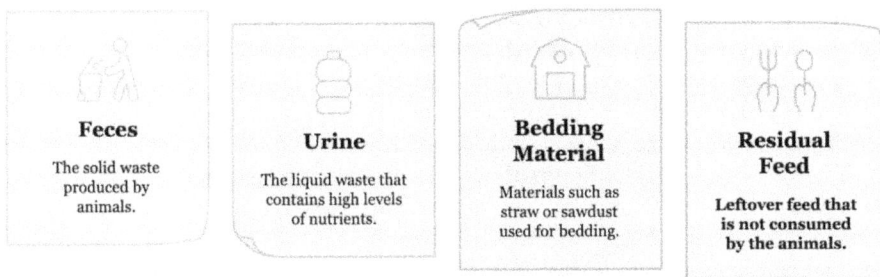

Feces

The solid waste produced by animals.

Urine

The liquid waste that contains high levels of nutrients.

Bedding Material

Materials such as straw or sawdust used for bedding.

Residual Feed

Leftover feed that is not consumed by the animals.

Figure 2.
Visual diagram illustrating the main components of animal manure, including feces, urine, bedding materials, and residual feed.

Waste type	Examples	Utilization methods
Fruit & Vegetable Waste	Peels, seeds, pulp	Composting, bioactive compound extraction
Cereal & Grain Waste	Bran, husks	Animal feed, biogas production
Dairy Waste	Whey, cheese residues	Protein extraction, probiotic production
Meat Processing Waste	Bones, blood, fat	Biodiesel, gelatin, animal feed
Aquaculture Waste	Fish scales, viscera, uneaten feed	Fishmeal, biogas, wastewater treatment

Table 1.
Major types of food processing and aquaculture waste with utilization methods.

through sustainable management strategies. Food processing waste can be composted to produce nutrient-rich organic fertilizers or converted into bioenergy through anaerobic digestion and fermentation. Additionally, biorefinery approaches allow for the extraction of valuable compounds such as antioxidants, dietary fibers, and proteins, which can be used in food, pharmaceuticals, and cosmetics. Waste oils and fats from food industries are utilized in biodiesel production, reducing dependency on fossil fuels [10].

Aquaculture waste is a significant source of organic nitrogen and phosphorus, which can cause water pollution if released, untreated. Sustainable solutions include wastewater treatment using constructed wetlands, integrated multi-trophic aquaculture (IMTA) where waste nutrients support secondary species like seaweed or shellfish, and biogas generation from fish processing residues. Fish by-products are also used in fishmeal and fish oil production, contributing to animal feed and nutraceutical applications. Adopting a circular economy approach to food and aquaculture waste minimizes environmental impact, promotes resource efficiency, and enhances economic sustainability. Effective policies, technological advancements, and industry collaboration are crucial for large-scale waste valorization. Major Types of Food Processing and Aquaculture Waste with Utilization Methods shown in **Table 1** [11].

3. Environmental and economic impacts of agricultural waste

3.1 Soil and water pollution

Improper management of agricultural waste significantly contributes to soil and water pollution, posing serious environmental and public health risks. Agricultural waste, including crop residues, animal manure, and food processing by-products, can introduce excess nutrients, heavy metals, pesticides, and organic contaminants into the environment, disrupting ecosystem balance and degrading natural resources. Soil pollution occurs when excessive organic or chemical waste accumulates in agricultural fields. The over-application of animal manure or untreated agricultural effluents leads to nutrient overload, resulting in soil acidification, salinization, and depletion of microbial diversity. The accumulation of heavy metals from fertilizers, pesticides, and wastewater irrigation can cause long-term soil contamination, reducing soil fertility and crop productivity. Additionally, the indiscriminate disposal of plastic mulch and packaging materials contributes to microplastic pollution, affecting soil structure and plant growth. Water pollution arises from the leaching and runoff of agricultural waste into surface and groundwater sources [12]. Excessive nitrogen and phosphorus from manure and fertilizer runoff promote eutrophication, leading to algal blooms, hypoxia, and loss of aquatic biodiversity. The infiltration of nitrates into groundwater is a major concern, as high nitrate levels in drinking water are linked to methemo-globinemia (blue baby syndrome) and other health disorders. Moreover, untreated wastewater from food processing industries and aquaculture systems can introduce pathogens, antibiotics, and organic pollutants into water bodies, endangering aquatic life and human health. Mitigation strategies to reduce soil and water pollution from agricultural waste include precision nutrient management, controlled application of organic amendments, and the adoption of buffer strips, constructed wetlands, and bioreactors to filter contaminants before they reach water bodies. Sustainable waste valorization techniques, such as composting, anaerobic digestion, and biochar appli-cation, help in reducing pollutant loads while improving soil health and water quality. Additionally, stringent environmental regulations and farmer education programs are essential for promoting responsible waste management practices, ensuring the protection of soil and water resources for sustainable agriculture [13].

3.2 Greenhouse gas emissions and climate change

Agricultural waste significantly contributes to greenhouse gas (GHG) emissions, exacerbating climate change through the release of methane (CH_4), nitrous oxide (N_2O), and carbon dioxide (CO_2). These gases originate from various waste sources, including crop residues, animal manure, food processing by-products, and aquacul-ture waste, influencing global warming potential (GWP) due to their varying atmo-spheric lifetimes and radiative forcing effects. Methane emissions primarily result from anaerobic decomposition of organic agricultural waste, particularly in unman-aged manure storage systems, flooded rice fields, and food waste decomposition in landfills. CH_4 has a global warming potential 25 times higher than CO_2, making it a critical target for mitigation efforts. Anaerobic digestion of manure and organic waste can capture CH_4 for biogas production, reducing its direct emissions into the atmosphere. Nitrous oxide emissions arise from microbial nitrification and denitri-fication in soils treated with excessive manure or synthetic fertilizers. N_2O is nearly 300 times more potent than CO_2 in terms of global warming potential and remains in the atmosphere for over a century [14]. Improper waste application leads to nitrogen runoff, contributing to eutrophication and indirect N_2O emissions. Strategies such as precision nutrient management, biochar amendment, and cover cropping can help minimize N_2O emissions. Carbon dioxide emissions from agricultural waste occur through biomass burning, decomposition, and energy-intensive waste processing methods. Open-field burning of crop residues is a major source of CO_2, along with black carbon, which accelerates global warming and air pollution. Sustainable alter-natives include residue incorporation, biochar production, and bioenergy conversion, which enhance carbon sequestration and reduce fossil fuel dependence. To mitigate agricultural waste-driven climate impacts, integrated waste management strategies such as composting, anaerobic digestion, precision agriculture, and carbon farming must be adopted. Policy interventions, carbon credit incentives, and technologi-cal advancements in waste valorization and circular economy models are essential for reducing agricultural GHG emissions and promoting climate resilience in food systems [15].

3.3 Economic losses and resource inefficiencies

Inefficient agricultural waste management leads to substantial economic losses and resource inefficiencies, impacting farmers, industries, and global food systems. The mismanagement of crop residues, animal manure, food processing by-products, and aquaculture waste results in lost economic opportunities, increased production costs, and environmental damage that further strains financial resources. One of the primary economic losses arises from the underutilization of biomass resources. Crop residues that could be used for bioenergy, compost, or animal feed are often burned or discarded, leading to not only financial losses but also environmental degradation. Similarly, unprocessed animal manure and food waste contribute to methane emissions and soil pollution, rather than being repurposed as organic fertilizers or biogas feedstock. The lack of efficient waste recycling infrastructure prevents farmers from capitalizing on these valuable by-products. Resource inefficiencies also extend to nutrient losses in agricultural systems. Excess manure and chemical fertilizers, when improperly managed, lead to runoff and leaching, causing eutrophication and reducing soil productivity [16]. The overuse of fertilizers increases input costs for farmers while failing to optimize nutrient retention in soils. Additionally, water-intensive waste management practices in food processing industries and aquaculture operations contribute to excessive water consumption and contamination, further exacerbating economic inefficiencies. Addressing these losses requires circular economy approaches that transform waste into value-added products. Technologies such as anaerobic digestion, composting, and biochar production offer sustainable solutions by converting agricultural waste into energy, soil amendments, and industrial raw materials. Economic incentives, including government subsidies, carbon credits, and waste valorization policies, can encourage farmers and industries to adopt sustainable waste management strategies. By optimizing resource use and minimizing waste, agricultural systems can enhance profitability, reduce environmental impact, and contribute to a more sustainable and resilient food production system [17].

4. Strategies for sustainable agricultural waste management

4.1 Composting and vermicomposting

Composting and vermicomposting are sustainable and biologically driven processes that convert organic agricultural waste into nutrient-rich soil amendments. These techniques help manage waste efficiently while improving soil health, reducing dependence on chemical fertilizers, and mitigating environmental pollution. Composting is an aerobic microbial process where organic matter, such as crop residues, animal manure, and food processing waste, is decomposed into humus-like material under controlled conditions. This process involves microbial activity, where bacteria, fungi, and actinomycetes break down complex organic compounds into simpler forms. The composting process requires optimal conditions, including a carbon-to-nitrogen (C: N) ratio of 25:1 to 30:1, adequate moisture (40–60%), aeration, and a temperature range of 45–65°C for thermophilic decomposition. The end product, compost, enhances soil structure, increases water retention, and provides essential nutrients, promoting plant growth. Additionally, composting reduces greenhouse gas emissions by diverting organic waste from landfills, where it would otherwise decompose anaerobically, releasing methane (CH_4) [2]. Advanced composting techniques, such as windrow composting, aerated static piles, and in-vessel composting, improve efficiency and nutrient retention. Vermicomposting is an enhanced biological process that employs earthworms (Eisenia fetida, *Eudrilus eugeniae*, or *Perionyx excavatus*) to break down organic matter into a nutrient-rich, fine-textured material known as vermicompost. Earthworms consume organic waste, digest it, and excrete vermicasts, which are rich in plant-available nutrients, growth-promoting hormones, and beneficial microbes. The vermicomposting process requires optimal conditions, including a moisture content of 60–80%, a temperature range of 15–30°C, and a balanced C:N ratio. Compared to traditional composting, vermicomposting produces high-quality organic fertilizer faster, with enhanced microbial diversity and improved nutrient availability. The final product increases soil aeration, suppresses plant pathogens, and enhances root development. Integrating composting and vermicomposting in agricultural systems reduces waste accumulation, enhances soil fertility, and lowers input costs. These eco-friendly approaches contribute to sustainable farming practices by closing nutrient loops and promoting circular economy principles. Governments and agricultural policies supporting on-farm composting, subsidies for composting units, and farmer training programs can further encourage widespread adoption. Combining composting with other waste valorization strategies, such as anaerobic digestion and biochar application, enhances overall resource efficiency in modern agriculture [18].

4.2 Biogas and biochar production

Biogas and biochar production are innovative and sustainable waste management strategies that convert agricultural residues into valuable bioenergy and soil-enhancing materials. These processes not only reduce waste accumulation but also contribute to circular economy principles by improving resource efficiency, reducing greenhouse gas emissions, and enhancing soil health. Biogas is produced through anaerobic digestion, a microbial process that breaks down organic waste in the absence of oxygen. Agricultural residues such as animal manure, crop residues, and food processing waste serve as feedstock for anaerobic digestion. The process involves four key stages: hydrolysis, acidogenesis, acetogenesis, and methanogenesis, leading to the production of biogas, primarily composed of methane (CH_4) and carbon dioxide (CO_2) [19]. Optimal conditions for biogas production include a temperature range of 35–55°C, a pH of 6.5–7.5, and an appropriate carbon-to-nitrogen (C:N) ratio. The biogas generated can be used as a renewable energy source for electricity generation, heating, and cooking, reducing reliance on fossil fuels. The by-product of anaerobic digestion, known as digestate, is rich in nutrients and can be used as an organic fertilizer to enhance soil fertility. Implementing biogas technology in agricultural systems can significantly reduce methane emissions from unmanaged manure and organic waste while providing energy security for rural communities. Biochar is a stable carbon-rich material produced through pyrolysis, a thermal decomposition process that occurs under limited oxygen conditions. Various agricultural wastes, including crop residues, wood chips, and animal manure, can be used as feedstock for biochar production [20]. The pyrolysis process operates at temperatures ranging from 300 to 700°C, resulting in the formation of biochar along with syngas and bio-oil as additional by-products. The application of biochar in agriculture improves soil structure, enhances moisture retention, and increases soil microbial activity. Biochar also acts as a carbon sequestration tool, capturing atmospheric carbon and storing it in the soil for extended periods, thereby mitigating climate change. Additionally, biochar enhances the efficiency of fertilizers by reducing nutrient leaching and improving plant nutrient uptake. Combining biogas and biochar production in agricultural waste management provides a dual benefit of renewable energy generation and soil improvement. Governments and research institutions are increasingly promoting these technologies through policy incentives, funding for bioenergy projects, and awareness campaigns. Widespread adoption of biogas and biochar technologies can enhance agricultural sustainability, reduce environmental pollution, and contribute to a more resilient global food system [21].

4.3 Phytoremediation and biological waste management

Phytoremediation and biological waste management are nature-based approaches that offer sustainable solutions for treating agricultural waste and mitigating environmental pollution. These methods harness the ability of plants and microorganisms to absorb, degrade, or transform contaminants, thereby reducing waste accumulation while enhancing soil and water quality. The integration of phytoremediation and biological processes can significantly contribute to agricultural sustainability by minimizing chemical inputs, improving resource efficiency, and restoring degraded ecosystems. Phytoremediation is a plant-based remediation technique that utilizes specific plant species to extract, stabilize, or degrade pollutants from soil and water [22].

Hyperaccumulator plants such as *Brassica juncea* (Indian mustard) and *Helianthus annuus* (sunflower) are known for their ability to absorb heavy metals and excess nutrients, preventing their accumulation in the environment. The process involves multiple mechanisms, including phytoextraction, where plants uptake contaminants and store them in their biomass, and phytostabilization, which immobilizes harmful substances within the root zone, reducing leaching into groundwater. Additionally, rhizofiltration employs plant roots to filter out pollutants from wastewater, while phytodegradation enables the breakdown of organic contaminants, such as pesticides and hydrocarbons, into non-toxic forms. Phytoremediation is a cost-effective and eco-friendly alternative to chemical and physical remediation techniques, although its efficiency depends on plant species, soil conditions, and pollutant type. Biological waste management involves the use of microorganisms and enzymes to degrade and recycle agricultural waste into valuable by-products. Microbial decomposition plays a crucial role in the breakdown of crop residues, animal manure, and food processing waste, converting organic matter into nutrient-rich compost or biofertilizers [23]. Specific bacteria, such as *Bacillus* and *Pseudomonas*, are effective in biodegradation, while fungi like *Trichoderma* contribute to composting and mycoremediation by breaking down complex organic molecules and toxic compounds. Mycoremediation, a fungal-based strategy, has shown potential in detoxifying pesticide-contaminated soils and enhancing nutrient cycling. Additionally, microbial fuel cells (MFCs) utilize bacteria to convert organic waste into bioelectricity, offering an innovative approach to energy recovery from agricultural residues. The integration of phytoremediation and biological waste management provides a holistic solution to agricultural waste challenges by promoting waste recycling, reducing soil and water contamination, and improving resource efficiency. Scaling up these technologies requires policy support, farmer education, and advancements in microbial and plant biotechnology. Future research should focus on enhancing plant-microbe interactions and optimizing biological degradation pathways to maximize their environmental and economic benefits. By adopting these nature-based solutions, agricultural systems can transition toward a more sustainable, circular economy with minimal environmental impact [24].

5. Technological innovations in waste management

5.1 Artificial intelligence (AI) and IoT in waste optimization

The application of Artificial Intelligence (AI) and the Internet of Things (IoT) in waste management represents a transformative shift toward more efficient and sustainable waste optimization. These technologies facilitate the automation, real-time monitoring, and data-driven decision-making necessary for improving waste-handling processes, enhancing resource recovery, and minimizing environmental impact. The convergence of AI and IoT holds the potential to optimize waste management at every stage, from waste generation to disposal, thus contributing to a circular economy framework. AI, particularly through machine learning (ML) algorithms, has demonstrated considerable potential in improving the operational efficiency of waste management systems. AI systems enable the processing of vast amounts of data generated by waste streams, allowing for predictive modeling of waste generation patterns, optimization of collection routes, and the identification of opportunities for recycling and resource recovery [25]. Machine vision systems powered by AI have been deployed in material recovery facilities (MRFs) to automate the sorting of waste based on characteristics such as shape, size, and composition. These AI-driven systems employ deep learning algorithms that significantly enhance the accuracy and speed of waste sorting, thereby reducing contamination and increasing recycling rates. AI is also being utilized in the optimization of waste collection. Through predictive analytics, AI can forecast waste generation trends, enabling more efficient allocation of resources and better planning for waste collection schedules. The implementation of AI-based routing algorithms for waste collection trucks reduces fuel consumption, minimizes traffic congestion, and decreases the carbon footprint of waste collection operations. The AI and IoT-driven smart waste management, optimizing collection, sorting, and recycling for a sustainable future, are shown in **Figure 3** [26]. IoT technology enhances waste management systems by providing continuous, real-time data collection from various components of the waste management process. Sensors embedded in waste containers monitor fill levels, temperature, and other environmental parameters, sending data to centralized systems. This real-time data allows waste management authorities to optimize collection routes, reduce unnecessary trips, and minimize operational costs. IoT-enabled sensors can also monitor the status of waste bins, preventing overflows and enabling timely intervention, which improves overall system efficiency. In addition, IoT systems are instrumental in tracking the movement of waste throughout its lifecycle, from collection to processing. By integrating IoT sensors in waste collection vehicles and treatment facilities, operators can gather critical data on the efficiency of waste processing, monitor equipment conditions, and perform predictive maintenance, thereby extending the lifespan of infrastructure and reducing downtime [27]. The integration of AI and IoT creates a powerful synergy that enhances waste management capabilities. IoT devices collect real-time data from various points within the waste management system, including waste bins, collection trucks, and processing facilities. This data serves as a foundation for AI-based analytics, which can detect patterns, predict waste generation trends, and provide recommendations for operational improvements. For example, AI can use IoT data to optimize bin placement strategies, adjust collection schedules dynamically based on real-time fill levels, and enhance recycling efficiency by detecting contamination or mis-sorting in real-time. Furthermore, AI and IoT integration enables the creation of smart waste management systems within urban

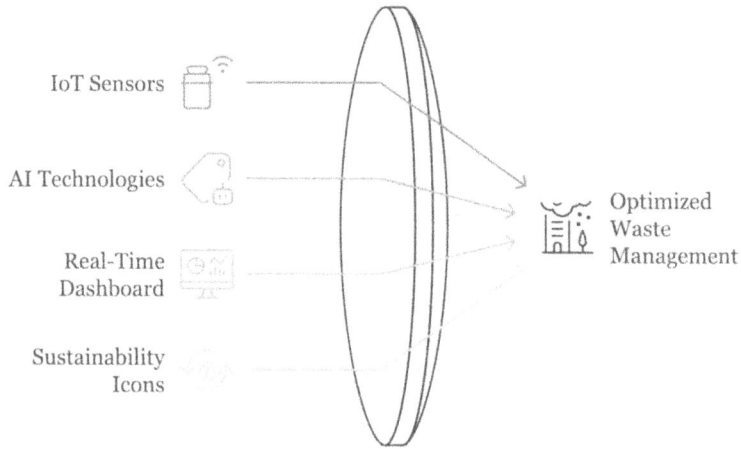

Figure 3.
The integration of AI and IoT in smart waste management, enhancing efficiency in collection, sorting, and recycling for sustainability.

settings, providing dynamic responses to fluctuating waste volumes and disposal needs. In smart cities, these technologies work in tandem to support decision-making, enabling policymakers to track waste diversion progress, evaluate the success of recycling initiatives, and make data-driven decisions on waste reduction strategies [28].

5.2 Biotechnology and precision agriculture

Biotechnology and precision agriculture have emerged as transformative approaches in modern farming, enhancing crop productivity, resource efficiency, and environmental sustainability. By integrating biotechnological advancements with data-driven precision techniques, these innovations address global challenges such as food security, climate change, and soil degradation. Biotechnology plays a crucial role in improving agricultural practices through genetic modification, microbial applications, and bioengineered solutions. Genetically modified (GM) crops, developed through recombinant DNA technology, exhibit traits such as pest resistance, herbicide tolerance, and improved nutritional content. These advancements reduce the reliance on chemical pesticides and fertilizers, promoting environmentally sustainable farming. Microbial biotechnology is another key area, utilizing plant growth-promoting rhizobacteria (PGPR) and mycorrhizal fungi to enhance nutrient uptake and soil fertility [29]. Biofertilizers and biopesticides derived from beneficial microorganisms provide eco-friendly alternatives to synthetic agrochemicals, minimizing environmental pollution and promoting soil health. Additionally, advancements in synthetic biology and CRISPR-based genome editing offer precise genetic modifications, enabling the development of resilient crop varieties with improved yield and stress tolerance. The integration of biotechnology and precision agriculture, showcasing genetic advancements, smart farming technologies, and their synergy for sustainable and efficient crop production, is shown in **Figure 4**. Precision agriculture employs advanced technologies such as remote sensing, geographic information systems (GIS), drones, and artificial intelligence (AI) to optimize agricultural inputs and improve efficiency. Real-time data collection from soil sensors, satellite imagery, and unmanned aerial vehicles (UAVs) allows farmers to make informed

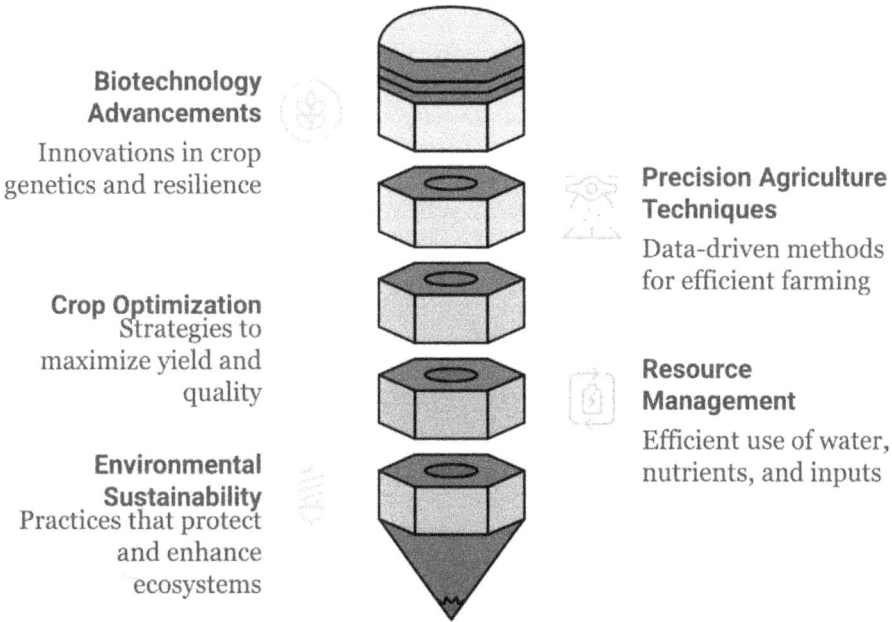

Figure 4.
Biotechnology and precision agriculture working together for sustainable crop production.

decisions on irrigation, fertilization, and pest control. Variable Rate Technology (VRT) enables site-specific application of water, nutrients, and pesticides, reducing wastage and enhancing resource efficiency. Automated machinery and AI-driven analytics facilitate predictive modeling for disease outbreaks, crop growth patterns, and weather fluctuations, ensuring proactive farm management. The integration of Internet of Things (IoT) devices in smart farming provides continuous monitoring of soil moisture, temperature, and crop health, allowing real-time adjustments to optimize productivity. This data-driven approach minimizes resource consumption, mitigates climate-related risks, and supports sustainable agricultural practices. The convergence of biotechnology and precision agriculture enhances crop resilience and productivity while reducing the ecological footprint of farming. Biotechnological advancements provide improved crop varieties and bio-based solutions, while precision agriculture ensures their optimal management through real-time monitoring and predictive analytics. This synergy contributes to sustainable food production, efficient land use, and reduced environmental impact, supporting global efforts toward climate-smart agriculture [30].

6. Circular economy in agriculture

6.1 Waste-to-resource approaches in agriculture

The transition to a circular economy in agriculture emphasizes the sustainable utilization of agricultural waste, transforming it into valuable resources. Waste-to-resource approaches focus on reducing environmental pollution, enhancing soil fertility, and promoting resource efficiency by recycling organic waste into biofertilizers, bioenergy, and value-added products. These strategies align with sustainable agricultural practices, minimizing reliance on synthetic inputs and reducing the carbon footprint of farming operations [31]. One of the most effective waste-to-resource strategies in agriculture is composting, which converts organic farm waste, such as crop residues, manure, and food waste, into nutrient-rich compost. This process enhances soil organic matter, improves microbial diversity, and reduces the need for chemical fertilizers. Vermicomposting, involving earthworms, accelerates decomposition and enhances the bioavailability of nutrients. These practices contribute to soil health restoration and carbon sequestration. Microbial biofertilizers derived from agricultural waste offer an eco-friendly alternative to synthetic fertilizers. Bacteria such as *Rhizobium*, *Azotobacter*, and phosphate-solubilizing microbes play a crucial role in nitrogen fixation and phosphorus mobilization, improving plant nutrient uptake. Similarly, biostimulants extracted from plant and microbial sources enhance plant growth and stress tolerance [32]. The reuse of agricultural residues to develop these bio-based solutions supports sustainable nutrient cycling. Biogas production through anaerobic digestion of livestock manure, crop residues, and food waste provides a renewable energy source while reducing methane emissions from organic waste decomposition. The digestate, a by-product of biogas production, serves as a nutrient-rich biofertilizer. Additionally, pyrolysis and gasification technologies convert agricultural biomass into biochar and biofuels, improving soil carbon storage and providing alternative energy sources for rural communities. Several agricultural by-products can be repurposed into value-added products. Crop residues, such as rice husks and wheat straw, are used for bio-based packaging, animal feed, and construction materials. Fruit and vegetable waste serve as raw materials for bioactive compounds, natural dyes, and biodegradable plastics. The extraction of high-value biochemicals from agro-waste further strengthens circular bioeconomy initiatives [33].

6.2 Case studies on circular economy models in agriculture

The circular economy in agriculture focuses on minimizing waste and maximizing resource efficiency by transforming agricultural by-products into valuable inputs. Various real-world implementations demonstrate how circular strategies contribute to sustainable farming. This section presents two case studies that highlight effective circular economy models: anaerobic digestion for biogas production and rice husk valorization [34]. The Netherlands has successfully integrated anaerobic digestion (AD) into dairy and livestock farming to convert organic waste into renewable energy and biofertilizers. AD is a microbial process that breaks down manure, crop residues, and food waste in oxygen-free conditions, producing biogas rich in methane. This biogas serves as an alternative energy source for farms, reducing dependence on fossil fuels. The digestate, a by-product of the process, is a nutrient-rich material that can be used as organic fertilizer to enhance soil fertility. Implementing AD in Dutch farms has significantly reduced greenhouse gas emissions by capturing methane, a potent climate-warming gas, and utilizing it for energy generation. Additionally, the use of digestate as a biofertilizer minimizes the need for chemical fertilizers, reducing nitrate runoff into water bodies. The circular approach not only promotes energy self-sufficiency but also closes nutrient loops, improving soil health and supporting sustainable crop production [35]. The success of AD in the Netherlands serves as a model for integrating waste-to-energy technologies in livestock and mixed farming systems worldwide. India, one of the largest rice producers, generates millions of tons of rice husks annually, posing a significant waste management challenge. Instead of discarding this agricultural by-product, innovative circular economy models have been developed to convert rice husks into biochar and other value-added products. Biochar, produced through pyrolysis, is a carbon-rich material that enhances soil structure, improves water retention, and sequesters atmospheric carbon, making it a key component in climate-resilient agriculture. In addition to biochar, rice husks are also repurposed as raw materials for bio-based packaging, insulation panels, and construction materials, reducing reliance on non-renewable resources. This approach not only mitigates agricultural waste but also generates economic opportunities for farmers and industries involved in bio-based product development. By integrating waste valorization into the agricultural system, India demonstrates how circular economy principles can contribute to sustainability, waste reduction, and economic growth.

7. Policy frameworks and regulatory aspects in circular agriculture

The transition to a circular economy in agriculture requires a robust policy framework that includes global and national regulations, incentive programs, and policy support. Effective governance ensures the sustainable management of agricultural resources, promotes waste reduction, and facilitates the adoption of circular economy models. Global organizations and national governments play a crucial role in shaping regulatory frameworks that drive sustainability in agriculture. This section explores key regulations and incentive mechanisms that support circular agriculture [36].

7.1 Global and National Regulations

International regulations and national policies serve as the foundation for implementing circular economy principles in agriculture. Global organizations, including the United Nations (UN), the Food and Agriculture Organization (FAO), and the European Union (EU), have established frameworks that encourage sustainable agricultural practices. One of the most significant global agreements is the UN Sustainable Development Goals (SDGs), particularly Goal 12 (Responsible Consumption and Production) and Goal 13 (Climate Action), which emphasize waste reduction, sustainable resource use, and climate-resilient agriculture. The FAO's Circular Economy Framework for Agriculture promotes sustainable food systems by advocating for waste valorization, bio-based solutions, and resource-efficient farming practices. At the regional level, the European Green Deal and the EU Circular Economy Action Plan mandate sustainable agricultural practices by enforcing regulations on organic waste recycling, bioenergy generation, and nutrient recovery. The Common Agricultural Policy (CAP) provides guidelines on resource efficiency, biodiversity conservation, and climate mitigation, integrating circular economy principles into agricultural policies. In the United States, the Food Recovery Act and the Agricultural Improvement Act (Farm Bill) support sustainable farming initiatives, including food waste reduction and bio-based product development. In India, the National Agroforestry Policy and the Solid Waste Management Rules, 2016 promote organic waste utilization, composting, and bio fertilizer production, ensuring efficient waste-to-resource transformation in agriculture. Similarly, China's Zero-Waste Agriculture Strategy emphasizes recycling agricultural residues and promoting bioenergy projects to minimize waste [37].

7.2 Incentive programs and policy support

To encourage the adoption of circular agriculture, governments provide financial incentives, subsidies, and technical support programs. These policies facilitate the transition to sustainable farming by reducing economic barriers and promoting the development of circular solutions. Many governments offer subsidies for bio fertilizers, composting units, and biogas plants to promote organic waste recycling. For example, the EU Horizon 2020 Program funds research and innovation projects in sustainable agriculture, supporting technological advancements in waste valorization and precision farming. The USDA Conservation Stewardship Program (CSP) provides financial assistance to farmers implementing regenerative agriculture practices, including nutrient cycling and composting. Tax incentives and grants also play a crucial role in circular agriculture. In India, the National Biogas and Organic Manure Program (NBOMP) provides financial support for setting up biogas plants, encouraging rural farmers to utilize livestock waste for renewable energy generation. China's Circular Agriculture Pilot Programs offer funding for bioenergy projects, eco-friendly fertilizers, and closed-loop farming systems. In addition to financial support, capacity-building initiatives and knowledge-sharing platforms strengthen policy implementation. The FAO and UNEP support training programs on sustainable agricultural practices, helping farmers adopt innovative waste-to-resource solutions. Public-private partnerships (PPPs) also facilitate the scaling of circular economy models, with industries investing in bio-based materials, sustainable packaging, and agro-waste processing technologies [38].

8. Challenges in waste management adoption

The transition to efficient waste management systems is essential for sustainability, yet various challenges hinder widespread adoption. These challenges can be broadly classified into economic, technological, socio-political, and awareness-related barriers. Addressing these issues requires a multi-stakeholder approach involving policymakers, industries, and communities to ensure the effective implementation of waste management solutions [39].

8.1 Economic and technological barriers

One of the major obstacles to adopting waste management practices is the high economic cost associated with advanced waste processing technologies. The establishment of waste treatment facilities, such as composting plants, biogas units, and recycling infrastructure, requires significant capital investment. Small and medium-sized enterprises (SMEs) and farmers, particularly in developing regions, often lack the financial resources to integrate sustainable waste management practices into their operations. Additionally, the cost of maintaining and operating these facilities can be a burden, discouraging long-term sustainability. Limited access to funding and subsidies further exacerbates this issue. Many developing countries lack strong financial support mechanisms for waste management initiatives, and private-sector investment in waste-to-resource projects remains insufficient. Government incentives and policy frameworks are often inadequate to attract investments, making large-scale waste management adoption difficult. Technological barriers also pose significant challenges. Many regions lack access to advanced waste processing technologies, resulting in inefficient waste disposal and pollution [40]. The integration of artificial intelligence (AI), the Internet of Things (IoT), and automated waste sorting systems requires skilled labor and infrastructure that are often unavailable in low-income areas. Additionally, research and innovation in biodegradable materials, waste-to-energy conversion, and precision waste management solutions are still in the development stage, limiting their practical application. Furthermore, inadequate waste collection and segregation systems hinder the effectiveness of recycling and reuse initiatives. In many urban and rural areas, waste is mixed at the source, making it difficult to separate organic and inorganic materials for appropriate processing. The absence of efficient waste segregation infrastructure leads to increased landfill dependency, exacerbating environmental pollution and resource wastage [41].

8.2 Socio-political and awareness challenges

Beyond economic and technological barriers, socio-political challenges significantly impact waste management adoption. One of the primary issues is the lack of well-defined policies and regulations governing waste management. In many countries, waste management laws are either poorly enforced or outdated, leading to inconsistencies in implementation. Corruption, bureaucratic inefficiencies, and inadequate governance further hinder progress, making it difficult to establish sustainable waste management frameworks. Political will and commitment play a crucial role in the successful adoption of waste management strategies. However, in many cases, governments prioritize short-term economic growth over long-term environmental sustainability. Industrial lobbies and resistance from stakeholders with vested interests in conventional waste disposal methods also delay the transition to circular economy models. The absence of clear incentives and legal mandates for industries and municipalities to adopt waste reduction strategies further slows progress [42].

Public awareness and behavioral challenges are equally critical. Many communities lack knowledge about sustainable waste disposal methods, leading to poor waste management practices. Inadequate education on the importance of waste segregation, recycling, and composting results in low participation in waste management programs. Cultural habits and resistance to change also contribute to ineffective waste disposal, particularly in regions where open dumping and burning of waste are common practices. Furthermore, consumer behavior significantly influences waste generation. The increasing consumption of single-use plastics, non-biodegradable packaging, and excessive resource exploitation contribute to waste accumulation. While some countries have introduced bans on plastic use and encouraged sustainable alternatives, widespread behavioral change is still lacking. Without proper education and incentives, consumer-driven waste reduction efforts remain limited. Another key challenge is the informal waste sector, which plays a significant role in waste collection and recycling in many developing nations. Informal waste pickers often operate in unsafe conditions with minimal income, lacking social security and legal recognition. Integrating this sector into formal waste management frameworks is essential for improving efficiency and sustainability. However, resistance from local authorities and limited policy support hinder such integration [43].

9. Future perspectives and recommendations in sustainable waste management

The transition toward sustainable waste management requires a comprehensive approach that integrates technological advancements, policy support, and behavioral changes. While current challenges hinder widespread adoption, scaling up sustainable practices, addressing research gaps, and fostering future innovations can drive long-term progress. A strategic roadmap is essential to enhance waste management efficiency, promote circular economy principles, and ensure environmental sustainability. To achieve large-scale sustainability in waste management, a multi-faceted strategy involving infrastructure development, financial investment, policy implementation, and community participation is necessary. Governments must prioritize the expansion of waste treatment facilities, including composting plants, biogas units, and recycling centers, to accommodate increasing waste volumes. Strengthening waste collection and segregation systems through smart waste management technologies, such as artificial intelligence (AI) and the Internet of Things (IoT), can enhance efficiency and resource recovery. Financial support mechanisms, including subsidies, tax incentives, and public-private partnerships, are crucial in promoting sustainable waste management practices. Encouraging private-sector investment in waste-to-resource projects can accelerate technological advancements and infrastructure expansion. Additionally, integrating the informal waste sector into formal systems can improve waste collection, sorting, and recycling efficiency, ensuring both economic and social benefits. Community engagement and education play a vital role in scaling up sustainability efforts. Awareness campaigns, behavioral change programs, and training initiatives can encourage waste reduction, segregation, and responsible disposal. Governments and environmental organizations should collaborate to promote zero-waste initiatives, extended producer responsibility (EPR), and sustainable product designs to minimize waste generation at the source.

Despite progress in waste management technologies, several research gaps remain that hinder large-scale implementation. There is a need for advanced biodegradable materials to replace non-recyclable plastics and reduce environmental pollution. Research on microbial and enzymatic waste degradation methods can enhance organic waste treatment, leading to more efficient composting and biogas production. Additionally, developing cost-effective and scalable waste-to-energy conversion technologies is essential for resource recovery. Emerging innovations, such as AI-powered waste sorting systems, blockchain-based waste tracking, and nanotechnology applications in waste treatment, present promising solutions for improving waste management efficiency. Future research should focus on optimizing automation, machine learning, and IoT-based smart waste monitoring systems to enhance operational efficiency and minimize human intervention. Policy-driven innovation must also be emphasized, with governments investing in circular economy research, sustainable packaging alternatives, and eco-friendly waste treatment solutions. Strengthening international collaborations and knowledge-sharing platforms can accelerate the development and adoption of cutting-edge waste management technologies.

10. Conclusion

Sustainable waste management is critical for environmental protection, resource conservation, and circular economy transition. However, several barriers hinder its large-scale implementation. Economic challenges include high capital investment, inadequate financial incentives, and limited private-sector participation. Technological barriers, such as inefficient waste segregation, inadequate recycling infrastructure, and restricted access to advanced treatment methods, further limit progress. Additionally, weak regulatory enforcement, fragmented policies, and

Agricultural Waste Management for Food Security and Sustainability
DOI: http://dx.doi.org/10.5772/intechopen.1010261

socio-political challenges, including low public awareness and resistance to behavioral change, pose significant obstacles. Despite these challenges, emerging technologies such as artificial intelligence (AI), the Internet of Things (IoT), and biotechnology offer promising solutions. AI-powered waste sorting systems, bio-based waste treatment methods, and IoT-enabled smart waste management enhance efficiency and promote resource recovery. Circular economy models, such as waste-to-energy conversion, composting, and biofertilizer production, present viable alternatives to conventional disposal practices. However, weak policy enforcement and limited investment in research and development slow their large-scale adoption. A multi-faceted approach integrating technological advancements, policy interventions, and financial mechanisms is essential for sustainable waste management. Governments must strengthen regulatory frameworks, enforce stricter waste management policies, and incentivize industries to adopt sustainable practices. Financial support mechanisms, including public-private partnerships (PPPs) and extended producer responsibility (EPR) programs, can accelerate waste management advancements and infrastructure development. Investment in innovative waste treatment technologies, such as anaerobic digestion, AI-driven waste sorting, and biotechnological solutions, is crucial for improving efficiency and scalability. Integrating informal waste collectors into formal systems can enhance waste segregation and recycling while ensuring social equity. Additionally, fostering interdisciplinary research in biodegradable materials, nanotechnology-based waste treatment, and block chain-enabled waste tracking can further optimize waste valorization. Public awareness and education campaigns are equally vital to promoting responsible waste disposal and sustainable consumption. Encouraging industries to adopt eco-friendly designs, sustainable packaging, and circular economy principles will further minimize waste generation. By combining technological innovation, policy enforcement, and community participation, sustainable waste management can be effectively scaled up. Strengthening international collaborations and knowledge-sharing platforms will accelerate the transition to circular economy models, ensuring long-term environmental sustainability and resource efficiency.

Acknowledgements

Authors and co-authors are grateful to Prof. P.B. Sharma, Vice Chancellor, Amity University Haryana, Gurugram, for the necessary help and support related to the completion of research work.

Funding

There is no funding available for the review article/research work related to design of the study and collection, analysis and interpretation of data including writing the Ms.

Declaration of interests

The authors declare that they have no known competing financial interests or personal relationships that could have appeared to influence the work reported in this chapter.

Consent for publication

It is not applicable as we have not incorporated data/figures/tables from any person or publishers. The figures and table are prepared by the authors of the Ms.

Competing interests

I like to say that authors co-authors do not declares that they have no competing interests. There is no financial interest. They are interested to publish "Agricultural Waste Management for Food Security and Sustainability" for publication in your esteemed Journal.

Availability of data and materials

The data and materials presented in the manuscript or additional supporting files and formats has not been taken from publicly available repositories. All the authors agreed to provide data and materials after request.

Credit authorship contribution statement

Krishan Kumar —Writing-original draft, conceptualization, investigation. Annu Khatri - investigation, writing, data curation, corrected the Ms. properly. Indu Shekhar Thakur- carried out initial planning, conceptualization, supervision, visualization, writing, editing, and reviewing.

Author details

Krishan Kumar, Annu Khatri and Indu Shekhar Thakur*
Amity School of Earth and Environmental Sciences, Amity University Haryana, Gurugram, Haryana, India

*Address all correspondence to: isthakur@hotmail.com and isthakur@ggn.amity.edu

IntechOpen

References

[1] Alan H, Köker AR. Analyzing and mapping agricultural waste recycling research: An integrative review for conceptual framework and future directions. Resources Policy. 2023;**85**:103987

[2] Sinha AK, Rakesh S, Mitra B, Roy N, Sahoo S, Saha BN, et al. Agricultural waste management policies and programme for environment and nutritional security. In: Input Use Efficiency for Food and Environmental Security. 2021. pp. 627-664

[3] Yates J, Deeney M, Rolker HB, White H, Kalamatianou S, Kadiyala S. A systematic scoping review of environmental, food security and health impacts of food system plastics. Nature Food. 2021;**2**(2):80-87

[4] Lin AYC, Huang STY, Wahlqvist ML. Waste management to improve food safety and security for health advancement. Asia Pacific Journal of Clinical Nutrition. 2009;**18**(4):538-545

[5] Awogbemi O, Von Kallon DV. Pretreatment techniques for agricultural waste. Case Studies in Chemical and Environmental Engineering. 2022;**6**:100229

[6] Kumar Sarangi P, Subudhi S, Bhatia L, Saha K, Mudgil D, Prasad Shadangi K, et al. Utilization of agricultural waste biomass and recycling toward circular bioeconomy. Environmental Science and Pollution Research. 2023;**30**(4):8526-8539

[7] Awasthi MK, Sindhu R, Sirohi R, Kumar V, Ahluwalia V, Binod P, et al. Agricultural waste biorefinery development towards circular bioeconomy. Renewable and Sustainable Energy Reviews. 2022;**158**:112122

[8] Mengqi Z, Shi A, Ajmal M, Ye L, Awais M. Comprehensive review on agricultural waste utilization and high-temperature fermentation and composting. Biomass Conversion and Biorefinery. 2021;**13**(7):5445-5468

[9] Koul B, Yakoob M, Shah MP. Agricultural waste management strategies for environmental sustainability. Environmental Research. 2022;**206**:112285

[10] Hasan Z, Lateef M. Transforming food waste into animal feeds: An in-depth overview of conversion technologies and environmental benefits. Environmental Science and Pollution Research. 2024;**31**(12):17951-17963

[11] Sikiru S, Abioye KJ, Adedayo HB, Adebukola SY, Soleimani H, Anar M. Technology projection in biofuel production using agricultural waste materials as a source of energy sustainability: A comprehensive review. Renewable and Sustainable Energy Reviews. 2024;**200**:114535

[12] Kumar M, Ambika S, Hassani A, Nidheesh PV. Waste to catalyst: Role of agricultural waste in water and wastewater treatment. Science of the Total Environment. 2023;**858**:159762

[13] Mishra S, Kumar R, Kumar M. Use of treated sewage or wastewater as an irrigation water for agricultural purposes-environmental, health, and economic impacts. Total Environment Research Themes. 2023;**6**:100051

[14] Sovacool BK, Griffiths S, Kim J, Bazilian M. Climate change and industrial F-gases: A critical and systematic review of developments, sociotechnical systems and policy options

for reducing synthetic greenhouse gas emissions. Renewable and Sustainable Energy Reviews. 2021;**141**:110759

[15] Bhatti UA, Bhatti MA, Tang H, Syam MS, Awwad EM, Sharaf M, et al. Global production patterns: Understanding the relationship between greenhouse gas emissions, agriculture greening and climate variability. Environmental Research. 2024;**245**:118049

[16] Bherwani H, Nair M, Niwalkar A, Balachandran D, Kumar R. Application of circular economy framework for reducing the impacts of climate change: A case study from India on the evaluation of carbon and materials footprint nexus. Energy Nexus. 2022;**5**:100047

[17] Maraveas C, Karavas CS, Loukatos D, Bartzanas T, Arvanitis KG, Symeonaki E. Agricultural greenhouses: Resource management technologies and perspectives for zero greenhouse gas emissions. Agriculture. 2023;**13**(7):1464

[18] Singh VK, Solanki P, Ghosh A, Pal A. Solid waste management and policies toward sustainable agriculture. In: Handbook of Solid Waste Management: Sustainability Through Circular Economy. Singapore: Springer Nature Singapore; 2022. pp. 523-544

[19] Kumar M, Dutta S, You S, Luo G, Zhang S, Show PL, et al. A critical review on biochar for enhancing biogas production from anaerobic digestion of food waste and sludge. Journal of Cleaner Production. 2021;**305**:127143

[20] Mensah-Sackey G, Shokry H, Fujii M, Nasr M. Sustainable utilization of plastic-derived graphene for tetracycline wastewater treatment and its recycling for biogas and biochar production. Journal of Water Process Engineering. 2025;**69**:106554

[21] Kalengyo RB, Ibrahim MG, Fujii M, Nasr M. Utilizing orange peel waste biomass in textile wastewater treatment and its recyclability for dual biogas and biochar production: A techno-economic sustainable approach. Biomass Conversion and Biorefinery. 2024;**14**(16):19875-19888

[22] Khan AHA, Kiyani A, Santiago-Herrera M, Ibánez J, Yousaf S, Iqbal M, et al. Sustainability of phytoremediation: Post-harvest stratagems and economic opportunities for the produced metals contaminated biomass. Journal of Environmental Management. 2023;**326**:116700

[23] Agrawal S, Kumar V, Singh S, Shahi SK. Gene mediated phytodetoxification of environmental pollutants. In: Phytoremediation Technology for the Removal of Heavy Metals and Other Contaminants from Soil and Water. Elsevier; 2022. pp. 405-433

[24] Yadav AN, Suyal DC, Kour D, Rajput VD, Rastegari AA, Singh J. Bioremediation and waste management for environmental sustainability. Journal of Applied Biology and Biotechnology. 2022;**10**(2):1-5

[25] Alsabt R, Alkhaldi W, Adenle YA, Alshuwaikhat HM. Optimizing waste management strategies through artificial intelligence and machine learning-an economic and environmental impact study. Cleaner Waste Systems. 2024;**8**:100158

[26] Pitakaso R, Srichok T, Khonjun S, Golinska-Dawson P, Sethanan K, Nanthasamroeng N, et al. Optimization-driven artificial intelligence-enhanced municipal waste classification system for disaster waste management. Engineering Applications of Artificial Intelligence. 2024;**133**:108614

[27] Pitakaso R, Srichok T, Khonjun S, Golinska-Dawson P, Gonwirat S, Nanthasamroeng N, et al. Artificial intelligence in enhancing sustainable practices for infectious municipal waste classification. Waste Management. 2024;**183**:87-100

[28] Shafik W. IoT-enabled model and waste management technologies for sustainable agriculture. In: IoT-Based Models for Sustainable Environmental Management: Sustainable Environmental Management. Springer Nature Switzerland: Cham; 2024. pp. 137-163

[29] Xing Y, Wang X. Precision agriculture and water conservation strategies for sustainable crop production in arid regions. Plants. 2024;**13**(22):3184

[30] Khan N. Unlocking innovation in crop resilience and productivity: Breakthroughs in biotechnology and sustainable farming. Innovation Discovery. 2024;**1**(4)

[31] Kolobaric A, Alagappan S, Hoffman L, Cozzolino D, Chapman J. From waste to resource: Managing plastics and persistent contaminants in the circular economy of food waste. Available at SSRN 5102114

[32] Patel K, Vashist M, Goyal D, Sarma R, Garg R, Singh SK. From waste to resource: A life cycle assessment of biochar from agricultural residue. Environmental Progress & Sustainable Energy. **44**:e14558

[33] Vijai, Wisetsri W. Circular economy for sustainable development in India. In: Global Sustainability: Trends, Challenges & Case Studies. Springer Nature Switzerland: Cham; 2024. pp. 61-87

[34] Cavicchi C, Oppi C, Vagnoni E. Energy management to foster circular economy business model for sustainable development in an agricultural

SME. Journal of Cleaner Production. 2022;**368**:133188

[35] Rekleitis G, Haralambous KJ, Loizidou M, Aravossis K. Utilization of agricultural and livestock waste in anaerobic digestion (AD): Applying the biorefinery concept in a circular economy. Energies. 2020;**13**(17):4428

[36] Kyriakopoulos GL. Circular economy and sustainable strategies: Theoretical framework, policies and regulation challenges, barriers, and enablers for water management. In: Water Management and Circular Economy. Elsevier; 2023. pp. 197-230

[37] Nicastro R, Papale M, Fusco GM, Capone A, Morrone B, Carillo P. Legal barriers in sustainable agriculture: Valorization of agri-food waste and pesticide use reduction. Sustainability. 2024;**16**(19):8677

[38] Velasco-Muñoz JF, Mendoza JMF, Aznar-Sánchez JA, Gallego-Schmid A. Circular economy implementation in the agricultural sector: Definition, strategies and indicators. Resources, Conservation and Recycling. 2021;**170**:105618

[39] Jalalipour H, Jaafarzadeh N, Morscheck G, Narra S, Nelles M. Adoption of sustainable solid waste management and treatment approaches: A case study of Iran. Waste Management & Research. 2021;**39**(7):975-984

[40] Bui TD, Tseng JW, Tseng ML, Lim MK. Opportunities and challenges for solid waste reuse and recycling in emerging economies: A hybrid analysis. Resources, Conservation and Recycling. 2022;**177**:105968

[41] Al-Emran M, Griffy-Brown C. The role of technology adoption in sustainable development: Overview,

opportunities, challenges, and future research agendas. Technology in Society. 2023;**73**:102240

[42] Tahiru AW, Cobbina SJ, Asare W. A circular economy approach to addressing waste management challenges in Tamale's waste management system. WORLD. 2024;**5**(3):659-682

[43] Anokye K. From waste to wealth: Exploring biochar's potential in energy generation and waste mitigation. Cleaner and Circular Bioeconomy. 2024:100101

Chapter 2

Waste to Wealth: The Circular Economy for Agricultural and Food Waste

Mudasir Ali and Hilaas Ahmad Peerzada

Abstract

In order to limit the consumption of natural resources without compromising economic growth, new economic models of production have been offered under the 12th Sustainable Development Goal "Responsible Consumption and Production" of the United Nations, which implies that, in addition to food losses, attention should be paid to the use and production of non-renewable resources and energy sources during the manufacturing and processing of food. In the agriculture and food sector, the circular economy is a paradigm that maximizes value from resources while minimizing waste by keeping them in use for as long as feasible. This strategy seeks to ensure that inputs and outputs are managed sustainably by closing the loop in food systems and agricultural production. Agriculture significantly contributes to global greenhouse gas emissions. Employing bioenergy in the agro-industrial sector can benefit from circular economy principles. Global initiatives have taken action to address the situation, even though only around 9% of the world's economy is circular. The circular economy (grow-make-use-restore) seeks to affect material and energy flows to enhance environmental benefits and minimize expenses, in contrast to the linear economy (take-make-use-dispose). In food systems, the circular economy entails lowering the quantity of waste produced within the system, reusing food, recycling nutrients, using food waste and by-products, and altering dietary habits to include more varied and effective feeding patterns. It is mostly possible to close the nutrition loop associated with the food system. Reducing food waste and surplus lowers the economic total matter consumption, which in turn lowers the linear economic matter flow.

Keywords: agricultural waste, bioenergy, food waste, circular economy, sustainable development goal 12

1. Introduction

These days, finding economical and ecologically friendly recycling techniques is essential due to serious environmental issues including soil and water pollution, taking sustainability and the circular economy (CE) into account with regard to food and plastic waste. Developing a society that generates less garbage can be considerably more successful than attempting to recycle it. Furthermore, it appears that less energy is needed for garbage formation as opposed to recycling procedures [1]. For example, in order to identify the root causes of this issue and to find a workable and sustainable solution to reduce food waste, researchers have looked into design concepts [2], user behaviour [3], habit theory [4], and discovery [5]. Additionally, it is known that food waste (FW) has an impact on the environment, society, and economy on a regional and worldwide scale [6, 7]. Food and green garbage make for 44% of the 2.01 billion tonnes of municipal solid trash produced annually worldwide [8]. As indicated in **Figure 1**, 37% of this amount is anticipated to be landfilled (only 8% of which have gas collection systems), 33% will be open disposed of, 19% will be recycled or composted, and 11% will be burned. The high rates of open dumping (33%) and landfilling without gas collecting equipment (29%) attest to the increased attention being paid to waste management [9]. Despite the substantial volume of waste generated, significant attention has been directed toward developing innovative and modern recycling techniques that are fast, clean, sustainable, and cost-effective. To achieve these benefits, the concept of a circular economy should be regarded as an essential strategy for effective waste management [10].

The circular economy is the sustainable way to turn these available wastes into valuable materials [11]. It has grown more and more important because of a number of causes, including the rise in trash, particularly every 5 years [12]. The shift from a linear to a circular economy is crucial because of the previously described factors [13]. There are numerous social, environmental, and economic advantages to adopting the circular model [14]. Widespread use of the economic cycle boosts the economy, which has several advantages, including more job opportunities, less raw material storage costs, less strain and negative environmental effects, and fewer price swings [15]. The international scientific and political community has recently begun to take a more active interest in circular economics, which aims to maintain products, materials, and components at the highest level of efficiency. To put it another way, the economic cycle's main objectives are to improve natural resource management and prevent waste creation by creating and ending material cycles [16, 17]. **Figure 2** shows the fundamentals of a circular economy.

The 12th Sustainable Development Goal (SDG 12) of the United Nations, titled "Responsible Consumption and Production," focuses on promoting sustainable consumption and production patterns globally. It aims to reduce the negative

Figure 1.
Global treatment and disposal of waste.

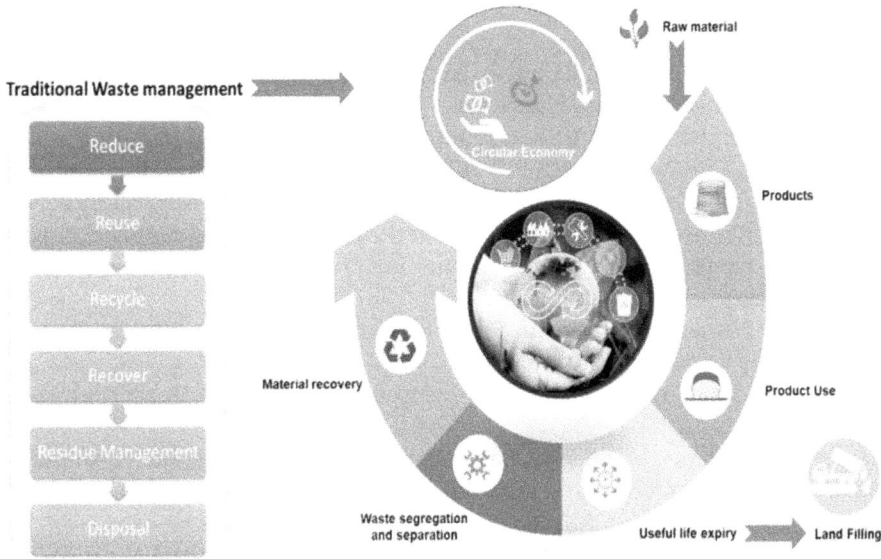

Figure 2.
Fundamentals of a circular economy.

environmental impacts associated with production and consumption, promote efficient use of resources, and minimize waste, while fostering economic growth and social development.

2. Key areas of focus in SDG 12: Responsible consumption and production

Efficient use of resources: SDG 12 seeks to encourage more economical use of natural resources. This entails cutting back on energy use, minimizing the use of non-renewable resources, and maximizing product lifecycles through circular economy models (reuse, recycling, and upcycling).

Waste reduction: Cutting back on waste production is one of SDG 12's main objectives. Reducing food waste, cutting back on single-use plastics, and stepping up recycling initiatives across industries are all part of this. Closing the loop is also necessary in a number of areas, including agricultural and food systems.

Sustainable supply chains: It is crucial to support industries in implementing sustainable production methods. This entails making use of eco-friendly designs, employing cleaner technology, and making sure that materials are sourced sustainably. Promotion of green technologies: SDG 12 advocates for the development and adoption of environmentally friendly technologies and practices that can reduce pollution, improve resource efficiency, and lower the carbon footprint.

Sustainable food systems: SDG 12 places a strong emphasis on minimizing food waste and losses, particularly in food systems. This entails promoting environmentally friendly, sustainable farming methods, guaranteeing food security, and streamlining the food supply system.

Consumer awareness: Educating customers about how their decisions affect the environment is another goal of SDG 12. This entails promoting sustainable products, lowering excessive consumption, and fostering responsible consumption habits.

2.1 Targets under SDG 12

The SDG 12 goal has specific targets that focus on tangible actions to address these issues. Some of the key targets [1] include:

Target 12.1: Implement the 10-Year Framework of Programs on Sustainable Consumption and Production (10YFP) and promote its adoption by all countries.

Target 12.2: Achieve effective and sustainable use of natural resources by 2030.

Target 12.3: Reduce food losses throughout the production and supply chains, including post-harvest losses, and cut the amount of food waste per person worldwide in half at the retail and consumer levels.

Target 12.4: Manage chemicals and their wastes throughout their life cycle in an environmentally responsible manner.

Target 12.5: Use prevention, reduction, recycling, and reuse to significantly lower waste production.

Target 12.6: Motivate businesses, particularly big and international businesses, to embrace sustainable practices and incorporate sustainability data into their reporting cycle.

Target 12.7: Encourage sustainable public procurement methods that align with national interests and policies.

Target 12.8: Making sure that people are aware of and have access to the information they need to live sustainably and in harmony with the environment.

2.2 Global action for SDG 12

While SDG 12 presents a universal framework, the implementation of responsible consumption and production requires collaboration across all sectors—government, businesses, consumers, and civil society. Various international initiatives and programs, such as the Ellen MacArthur Foundation for the circular economy, the Global Food Loss and Waste Initiative by the United Nations Environment Programme (UNEP), and the Sustainable Development Solutions Network (SDSN), work toward advancing this goal through research, partnerships, and policy advocacy.

2.2.1 Transition from a linear to a circular economy: An essential shift toward sustainability

Following the Second Industrial Revolution in 1850, when a linear production model predominately relied on natural resources for manufacturing and disposal, the planet has been experiencing an ecological catastrophe [18]. Due to this issue, there has been an impact or absence of 60% of the ecosystems on Earth. Furthermore, the world's population is using resources at a rate equal to 1.5 planets every year, exceeding the planet's ability to replenish resources and handle waste [18]. There were 8 billion people on Earth as of November 2022. According to projections, there will be 8.5 billion people on the planet by 2030, and by 2050, there will be 9.7 billion. The world's population is predicted to reach its high in 2080, when it will be 10.4 billion, and stay there until 2100 [19]. In order to shift from a linear production model to a CE, in which resources are continuously recycled and utilized rather than exploited and discarded, economists have been investigating potential solutions [20]. Designing out waste and pollution, preserving used goods and materials, and renewing natural systems are the three guiding concepts of the CE [21]. It is projected that this shift will lessen environmental effects while simultaneously generating economic benefits, such as the establishment of new business models and jobs [20].

Through international collaborations like the SDGs and, more recently, the idea of the CE, nations have been developing strategies in line with sustainable development [22, 23]. The goal of the CE is to lessen the linearity of systems of production and consumption, as well as the flow of energy and materials used in product creation. The CE encourages material reuse cycles rather than a linear production technique, which supports sustainable development [23]. The CE is showing promise as a means of fostering social and environmental justice while attaining a thriving economy within regional and global bounds. The CE is a complex idea with interrelated procedures rather than a straightforward fix. This intricacy highlights how difficult it is to evaluate it thoroughly [24]. According to Kirchherr et al. [25], the CE is in line with sustainability guidelines and serves as a tool to guide stakeholders in putting sustainable development ideas into practice. Nevertheless, despite conceptualizations, some CE research ignores the wider societal ramifications of a circular system in favor of concentrating only on social issues and sustainable development [26]. Additionally, it is acknowledged that more research on the CE at the macro level, such as the SDGs, is necessary. SDG 12 (responsible consumption and production) and CE practices are related, according to Schroeder et al. [27], particularly if they use intersectoral interaction models across the product chain life cycle. In this regard, productivity and efficiency measurement techniques like Data Envelopment Analysis (DEA) and the Malmquist Productivity Index (MPI) have become more popular, making it possible to create CE indicators. DEA is a nonparametric technique that compares inputs and outputs to assess the effectiveness of decision-making units, such as businesses or organizations, and shows how effectively resources are used in comparison to the units that perform the best. A technique for tracking changes in productivity over time is the Malmquist Productivity Index (MPI). It tracks improvements or decreases in performance by using DEA to evaluate changes in efficiency and technology.

Using DEA, a number of studies pertaining to SDG 12 (responsible consumption and production) have been found in the literature. SDG 12 requires examining environmental regulation concerns in order to achieve responsible production and consumption. In this regard, Gao et al. [28] looked into how environmental laws affected a country's development in carbon productivity. Stricter environmental regulations were found to be linked to higher overall carbon industrial productivity development in a country (e.g., China). Additionally, modernizing industrial structures and removing backwardness in the productive capacity of highly polluting industries are crucial steps in fostering positive synergies in carbon productivity.

2.3 Reduction of food waste and the circular economy

The development of a closed-loop process is one of the primary goals of the circular economy. In terms of food waste, this approach aims to improve the effectiveness of 9Rs-related processes and encourage a slowdown in flow, guaranteeing that food is utilized to its maximum capacity before being thrown away or disposed of [29, 30]. According to Borrello et al. [31], circular business models are successful in addressing societal issues and advancing the SDGs, particularly when it comes to advocating for a modern strategy that prioritizes sustainable industrial system measures and rethinks supply chain organization. Improving the distribution of food fit for human consumption or turning food waste into goods to make fertilizer, energy, biomaterials, etc., are two examples of food waste management strategies that include reducing food waste [31]. Recycling and the decrease in the production of these wastes may work along with the appropriate disposal of food waste through organic solid waste management programs, as suggested by the European Commission [32].

2.4 Strategies for waste reduction

According to Beckman et al. [33], the universal food system consumes a substantial amount of energy, accounting for almost 15–20% of overall energy use. With this level of energy usage, careful planning is necessary to make the most use of resources for coming generations. The main goal is to reduce waste output or prevent food from going to waste because doing so will require less energy to recycle, which can ultimately contribute to a cleaner environment. The entire Spanish waste management strategy for food packaging trash, from the collection stage of each part in specific packages to the ultimate recycling, employing eight different materials, was investigated. Additionally, specific life cycle impacts for each material at each stage were investigated. This research can serve as a foundation for studies comparing technologies and trash return systems, as well as assist in identifying current factors that greatly aid in the development of a unified strategy to control waste packaging. Reducing domestic food waste is one of the most successful waste reduction techniques that has drawn a lot of attention lately. In this instance, the key concept is to create a culture and focus on managing home garbage in order to reduce waste production. Additionally, focusing on household waste management may lead to positive improvements on a personal, social, and even international level [34]. Kim et al. [35] looked at different consumer viewpoints on reducing home food waste in Australia in a similar study. The research paths often employed in social marketing and proportion approaches to lower household food waste were presented to consumers in this study [35]. Nowadays, waste management of construction workers requires attention due to population increase and the rise of the building industry. In this context, Yang et al. looked into practical actions and elements that could lower worker-generated construction waste. This study's objectives were to determine the primary factors influencing construction workers' waste reduction practices using the theory of planned behavior (TPB) and to create a simulation model for those practices using a system dynamics approach (SD) [36].

2.5 Enhancing recycling rates

2.5.1 Institutional wastes

A significant amount of food is wasted annually, particularly in nations with less advanced science and technology. On the one hand, more research is needed on this topic, especially in universities, as they contribute to food waste because of their high student populations and abundant food consumption. Thus, a variety of recycling, sustainability, and assessment techniques can be used to look at this food waste at the university. Numerous food waste treatment technologies are available for this purpose, including composting [28], anaerobic digestion [37], thermal/incineration [38], landfilling [39], and protein recovery treatment [40]. Brenes-Peralta et al. conducted a similar study at five universities in Costa Rica and Latin America to list. The results showed that compared to valorization scenarios, the waste dump scenario had a higher potential for global warming and freshwater eutrophication. But there are additional expenses and impact divisions that are impacted [41].

Keng et al. conducted a life cycle estimate advanced case study on food waste composting (**Figures 3** and **4**) at the University of Nottingham Malaysia [42]. Furthermore, **Figure 5** illustrates that there are four primary processes in the recycling process (considering the most optimal states) [42]. Composting is one of the most environmentally friendly methods of managing food waste, taking into account all of the previously listed variables [43]. Studies on residential and commercial composting have been conducted, taking into account the compost's stability, chemical characteristics, and environmental impact. The rate at which degradable organic components break down is known as compost stability. Chemical metrics pertaining to its agronomic uses, such as organic material and nutritional content, did not differ significantly. Nonetheless, the rate of stability attained varied significantly. Compared to industrial compost, home compost can attain a higher degree of stability. To improve the stability of commercial composts, some composting operations need to put in a lot of work [44]. Waste sorting is a crucial component of food waste management that is unavoidable if a circular economy strategy is to be used to handle food waste as efficiently as possible. Modern actuators, sensors, control algorithms, and sorting procedures for recycling source-separated materials are only a few of the automated sorting systems that researchers are now carefully examining for effective waste management. Developed nations look for the best methods for recycling and removing waste from its source. However, source segregation is not taken into account in poor nations, typically as a result of low motivation or inadequate procedures like door-to-door collecting. This results in the mixed type of collected waste being disposed of in landfills without being separated, which is why automated waste sorting is necessary [45]. The results of this review indicate

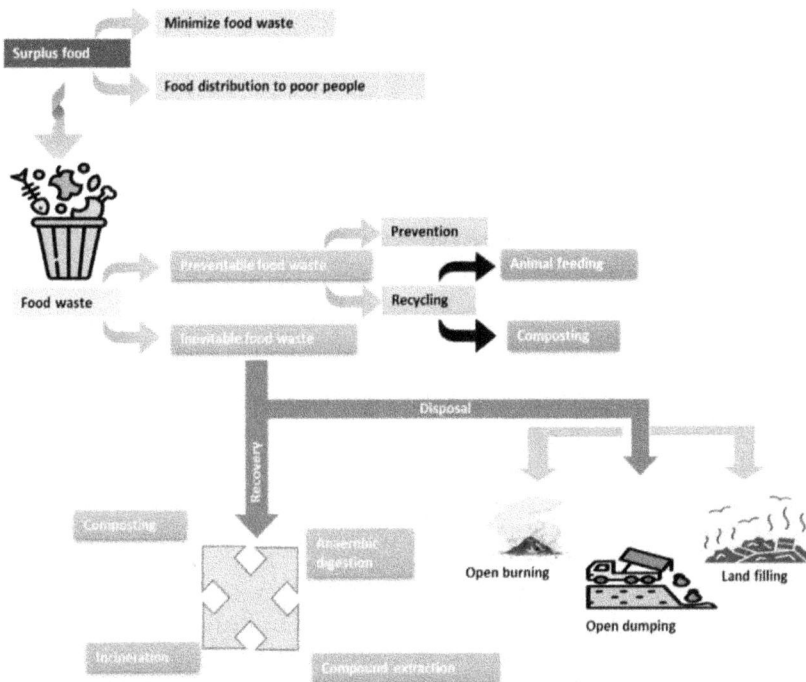

Figure 3.
The four stages for food waste management and treatment include reduce, reuse, recycle/recovery, and disposal.

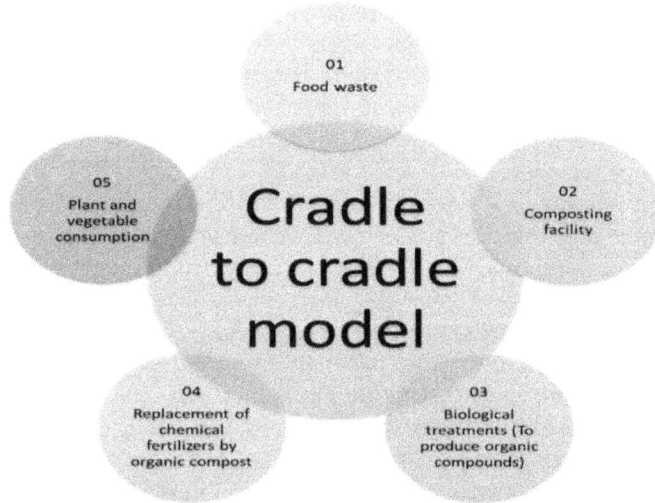

Figure 4.
Schematic diagram of cradle-to-cradle model steps.

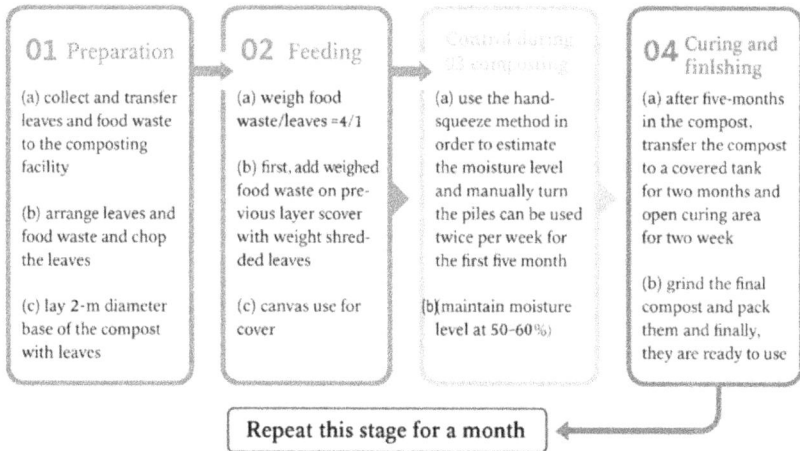

Figure 5.
The food waste treatment process.

that, as stressed by International Standards [46], efforts must be made to give food waste management scientific value and ensure that the profitability of the circular economy will be approached.

2.6 Wastes from agriculture

In order to generate social and economic benefits, new methods for the value-adding of industrial food leftovers must be developed to sustain agricultural and rural operations [47, 48]. Nowadays, many individuals worldwide are consuming more vegetables and agricultural goods as a result of the importance of a vegetarian diet. The Food and Agriculture Organization (FAO) of the United Nations reported that over 60% of trash comes from fruits and vegetables, whereas 45% of waste is associated with these materials. Without a doubt, this high consumption necessitates sophisticated and intelligent management for recycling and environmentally friendly waste generation. Many academics have recently become interested in powders made from fruit waste as a novel way to use these wastes [49–51]. Similarly, Bas-Bellver et al. created powders for use as food additives and functional food ingredients utilizing vegetable waste from a cooperative's fresh production lines (carrots, leeks, celery, and cabbage). Phenolic content rises dramatically during the process in comparison to unprocessed waste, especially when water is removed by air drying. This sequence is very clear for carrots and white cabbage, in contrast to leeks. On the other hand, green leeks and celery showed a drop in phenolic concentration over time. The findings were more nuanced when it came to flavonoid content [52]. Therefore, drying or conserving a small amount of water are appropriate technique to improve the outcomes and generate more stable powders. These powders are used in many food businesses as natural preservatives, flavorings, grain, and additives to boost the nutritional content of various goods. In a similar vein, Ortiz et al. [53] investigated the management of production waste in small orange water manufacturing enterprises. The authors of this study examined the technical, financial, and environmental aspects of burning and anaerobic digestion as viable garbage burial methods. The stability and value of these powders will be significantly increased if they are properly dried, according to the results [53].

2.7 Features and components of food waste

Food waste comes from a variety of sources, including fruit and vegetable waste from homes, restaurants, and marketplaces [54]. It is extremely diverse and contains substantial amounts of organic material and complicated constituents [55]. According to a number of studies, the organic composition of food waste varies, including protein, lipids, carbohydrates (including cellulose, hemicellulose, lignin, and starch), and organic acids [56, 54]. The availability and makeup of food waste can be impacted by variables like seasonality, market trends, geographic location, and sociocultural influences and behaviors (i.e., household food intake and expenditure) [57, 58]. Current food waste sources, as well as possible food waste paths and consequences, are depicted in **Figure 6**. The idea of a sustainable biorefinery for food waste that generates bioenergy and establishes local circular economies, also known as bioeconomies, is depicted in **Figure 7**. The development of a circular economy for food waste should prioritize decreasing the use of natural resources throughout the food supply chain system and valuing the trash, according to Read et al. [59]. For example, Ohja et al. [60] showed that using the circular economy to consider food waste for insect-based bioconversion could result in promising financial gains and commercialization as a source of high-value products (such as fertilizer, animal feed, and nutrients) and bioenergy (such as biodiesel). In order to add value to food waste as a bioenergy source, the government must support the development and adoption of a circular economy. Setting a vision at all levels, communicating with relevant parties, providing financial incentives, enlisting the aid of urban management levers, and enacting laws and regulations (i.e., consumer protection, product information, and consumer empowerment) are the five key government functions for implementing circular economy models in addressing food waste, according to a previous study [61].

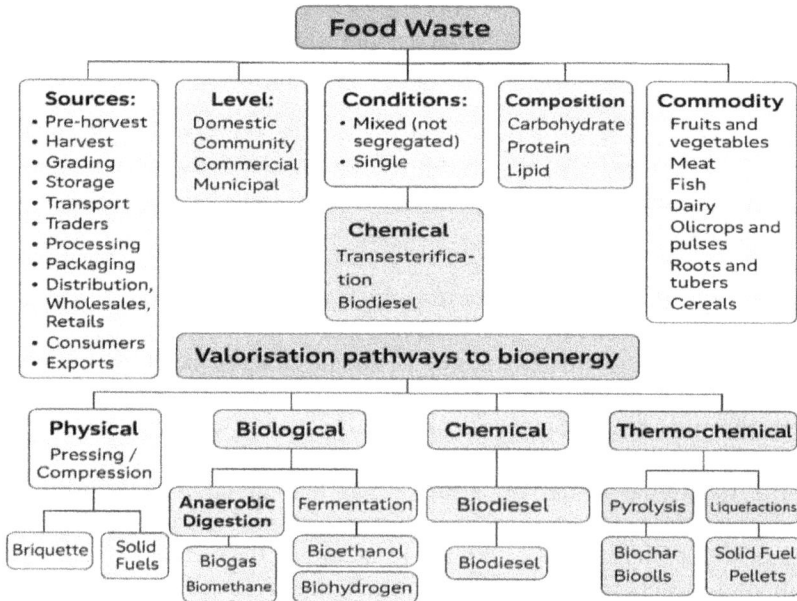

Figure 6.
Food waste valorization pathways for bioenergy production.

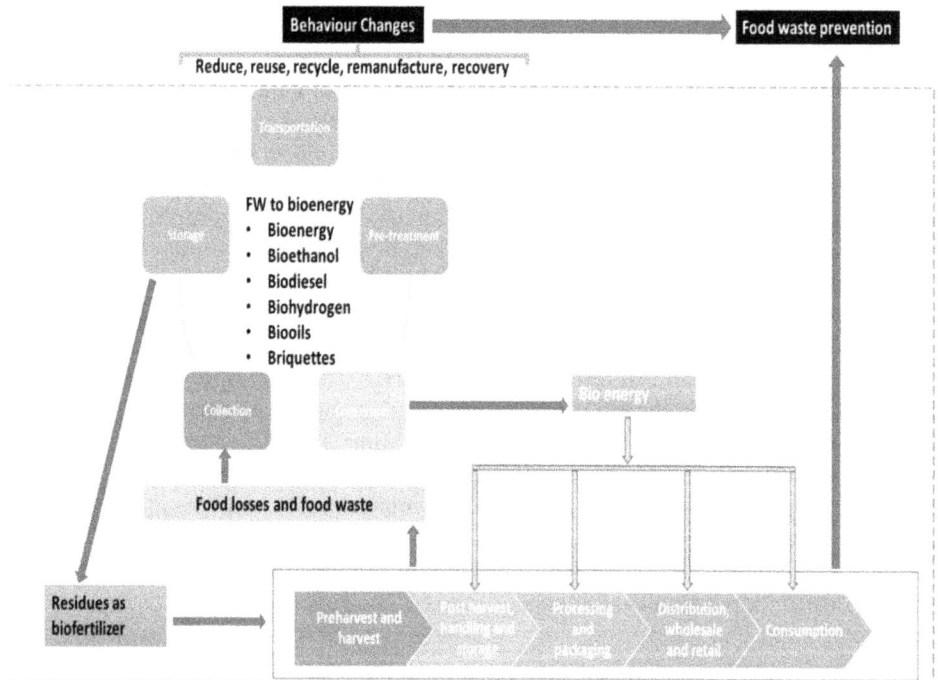

Figure 7.
Concept of circular economy for food waste to bioenergy.

3. Conclusion

In conclusion, the adoption of circular economy models and sustainable waste management practices is crucial for addressing environmental challenges. These strategies not only reduce waste but also promote resource efficiency, mitigate negative environmental impacts, and create economic opportunities. CE aligns with global sustainability goals, like SDG 12, which focuses on responsible production and consumption. A transition from a linear to a circular economy is essential to prevent ecological damage caused by resource depletion and waste, especially with a growing global population. CE promotes waste recycling, resource renewal, and energy efficiency, offering both environmental and economic benefits. However, implementing CE requires significant research and innovation. Food waste management plays a key role in CE, with strategies such as redistributing edible food, composting, and converting waste into valuable products like biofuels and fertilizers. Studies from universities and institutions emphasize the potential of composting, anaerobic digestion, and protein recovery technologies. Effective waste sorting and recycling are vital for establishing circular economies, with advanced systems being explored in developed nations, while developing countries face challenges with waste segregation. Additionally, agricultural and industrial food waste can be transformed into functional products, such as powders, animal feed, and biodiesel, through methods like drying, composting, and insect-based bioconversion. These processes not only reduce waste but also contribute to sustainability and economic growth. While there is potential for significant benefits from biorefinery strategies, including using food waste for bioenergy and valuable products, further research is needed to assess their viability. Overall, circular economy principles in food waste management offer a promising solution for a sustainable future.

Acknowledgements

Sincere gratitude is extended by the authors to the project staff of Division of Renewable Energy Engineering at the College of Agricultural Engineering and Technology, Sher-e-Kashmir University of Agricultural Sciences and Technology, for their assistance and contributions in the creation of this chapter.

Conflict of interest

The authors declare no conflict of interest.

Author details

Mudasir Ali* and Hilaas Ahmad Peerzada
Division of Renewable Energy Engineering, College of Agricultural Engineering and
Technology, Sher-e-Kashmir University of Agricultural Sciences and Technology,
Shalimar, Jammu and Kashmir, India

*Address all correspondence to: mudasir81fmp@gmail.com

IntechOpen

References

[1] United Nations. Transforming our World: The 2030 Agenda for Sustainable Development. New York, NY: United Nations. Available from: https://sustainabledevelopment.un.org/content/documents/21252030%20Agenda%20for%20Sustainable%20Development%20web.pdf; 2015 [Accessed: March 07, 2025]

[2] Hebrok M, Boks C. Household food waste: Drivers and potential intervention points for design – An extensive review. Journal of Cleaner Production. 2017;**151**:380-392

[3] Block LG et al. The squander sequence: Understanding food waste at each stage of the consumer decision-making process. Journal of Public Policy & Marketing. 2016;**35**:292-304

[4] Schanes K, Dobernig K, Gözet B. Food waste matters – A systematic review of household food waste practices and their policy implications. Journal of Cleaner Production. 2018;**182**:978-991

[5] Baron S, Patterson A, Maull R, Warnaby G. Feed people first: A service ecosystem perspective on innovative food waste reduction. Journal of Service Research. 2018;**21**:135-150

[6] Gentil EC, Poulsen TG. To Waste or Not to Waste – Food? London, UK: Sage Publications; 2012

[7] Thyberg KL, Tonjes DJ. The environmental impacts of alternative food waste treatment technologies in the US. Journal of Cleaner Production. 2017;**158**:101-108

[8] Kaza S, Yao L, Bhada-Tata P, VanWoerden F. What a Waste 2.0: A Global Snapshot of Solid Waste Management to 2050. Washington, DC: World Bank Publications; 2018a

[9] Kaza S, Yao LC, Bhada-Tata P, Van Woerden F. What a Waste 2.0: A Global Snapshot of Solid Waste Management to 2050. Washington, DC: Urban Development, The World Bank; 2018b. Available from: https://openknowledge.worldbank.org/handle/10986/30317

[10] Bigdeloo M, Teymourian T, Kowsari E, Ramakrishna S, Ehsani A. Sustainability and circular economy of food wastes: Waste reduction strategies, higher recycling methods, and improved valorization. Materials Circular Economy. 2021;**3**:1-9

[11] World Bank. What a Waste 2.0: A Global Snapshot of Solid Waste Management to 2050. License: CC BY 3.0 IGO. Washington, DC: World Bank Publications; 2018

[12] Dahiya S, Katakojwala R, Ramakrishna S, Mohan SV. Biobased products and life cycle assessment in the context of circular economy and sustainability. Materials Circular Economy. 2020;**2**:1-28

[13] Jose R, Panigrahi SK, Patil RA, Fernando Y, Ramakrishna S. Artificial intelligence-driven circular economy as a key enabler for sustainable energy management. Materials Circular Economy. 2020;**2**:1-7

[14] Mu'azu ND, Blaisi NI, Naji AA, Abdel-Magid IM, AlQahtany A. Food waste management current practices and sustainable future approaches: A Saudi Arabian perspectives. Journal of Material Cycles and Waste Management. 2019;**21**:678-690

[15] Maina S, Kachrimanidou V, Koutinas A. A roadmap towards a circular and sustainable bioeconomy through waste valorization. Current Opinion in Green and Sustainable Chemistry. 2017;**8**:18-23

[16] Kalmykova Y, Sadagopan M, Rosado L. Circular economy – From review of theories and practices to development of implementation tools. Resources, Conservation and Recycling. 2018;**135**:190-201

[17] Ingrao C, Faccilongo N, Di Gioia L, Messineo A. Food waste recovery into energy in a circular economy perspective: A comprehensive review of aspects related to plant operation and environmental assessment. Journal of Cleaner Production. 2018;**184**:869-892

[18] Zeller V, Towa E, Degrez M, Achten WM. Urban waste flows and their potential for a circular economy model at city-region level. Waste Management. 2019;**83**:83-94

[19] Ellen MacArthur Foundation. The Covid-19 Recovery Requires a Resilient Circular Economy. Cowes, UK: Ellen MacArthur Foundation; 2020. Available from: https://www. ellenmacarthurfoundation.org/articles/ the-covid-19-recovery-requires-a-resilient-circular-economy

[20] United Nations. World Population Dashboard. New York, NY: UNFPA; 2024. Available from: https://www.unfpa. org/data/world-population-dashboard

[21] United Nations. 2030 Agenda for Sustainable Development. New York, NY: United Nations; 2015. Available from: https://sustainabledevelopment.un.org/ content/documents/21252030%20 Agenda%20for%20Sustainable%20 Development%20web.pdf

[22] Ellen MacArthur Foundation. Explore the Circular Economy. Cowes, UK: Ellen MacArthur Foundation; 2022. Available from: https://www. ellenmacarthurfoundation.org/explore

[23] Korhonen J, Nuur C, Feldmann A, Birkie SE. Circular economy as an essentially contested concept. Journal of Cleaner Production. 2018;**175**: 544-552

[24] Van Bueren BJ et al. Comp-rehensiveness of circular economy assessments of regions: A systematic review at the macro-level. Environmental Research Letters. 2021;**16**(10):103001

[25] Kirchherr J, Reike D, Hekkert M. Conceptualizing the circular economy: An analysis of 114 definitions. Resources, Conservation and Recycling. 2017;**127**:221-232

[26] Merli R, Preziosi M, Acampora A. How do scholars approach the circular economy? A systematic literature review. Journal of Cleaner Production. 2018;**178**:703-722

[27] Schroeder P, Anggraeni K, Weber K. The relevance of circular economy practices to the sustainable development goals. Journal of Industrial Ecology. 2019;**23**(1):77-95

[28] Gao G, Wang K, Zhang C, Wei Y-M. Synergistic effects of environmental regulations on carbon productivity growth in China's major industrial sectors. Natural Hazards. 2019;**95**(1):55-72

[29] Lehtokunnas T, Mattila M, Närvänen E, Mesiranta N. Towards a circular economy in food consumption: Food waste reduction practices as ethical work. Journal of Consumer Culture. 2020;**22**(1):227-245. DOI: 10.1177/1469540520926252

[30] da Silva Duarte K, da Costa Lima TA, Alves LR, do Prado Rios PA, Motta WH. The circular economy approach for reducing food waste: A systematic review. Revista Produção e Desenvolvimento. 2021;7:1-15

[31] Borrello M, Pascucci S, Caracciolo F, Lombardi A, Cembalo L. Consumers are willing to participate in circular business models: A practice theory perspective to food provisioning. Journal of Cleaner Production. 2020;259:121013. DOI: 10.1016/j.jclepro.2020.121013

[32] Durrani K. Waste management and collaborative recycling: An SDG analysis for a circular economy. European Journal of Sustainable Development. 2019;8(5):197-197. DOI: 10.14207/ejsd.2019.v8n5p197

[33] Beckman J, Borchers A, Jones CA. Agriculture's supply and demand for energy and energy products. In: USDA-ERS, Economic Information Bulletin, No. 112. Washington, DC: United States Department of Agriculture, Economic Research Service; 2013

[34] Dietz T, Gardner GT, Gilligan J, Stern PC, Vandenbergh MP. Household actions can provide a behavioral wedge to rapidly reduce US carbon emissions. Proceedings of the National Academy of Sciences. 2009;106:18452-18456

[35] Kim J, Rundle-Thiele S, Knox K, Burke K, Bogomolova S. Consumer perspectives on household food waste reduction campaigns. Journal of Cleaner Production. 2020;243:118608

[36] Yang B, Song X, Yuan H, Zuo J. A model for investigating construction workers' waste reduction behaviors. Journal of Cleaner Production. 2020;265:121841

[37] Zhang C, Su H, Baeyens J, Tan T. Reviewing the anaerobic digestion of food waste for biogas production. Renewable and Sustainable Energy Reviews. 2014;38:383-392

[38] Elkhalifa S, Al-Ansari T, Mackey HR, McKay G. Food waste to biochars through pyrolysis: A review. Resources, Conservation & Recycling. 2019;144:310-320

[39] Ma P, Ke H, Lan J, Chen Y, He H. Field measurement of pore pressures and liquid-gas distribution using drilling and ERT in a high food waste content MSW landfill in Guangzhou, China. Engineering Geology. 2019;250:21-33

[40] Nguyen TT, Tomberlin JK, Vanlaerhoven S. Ability of black soldier fly (Diptera: Stratiomyidae) larvae to recycle food waste. Environmental Entomology. 2015;44:406-410. DOI: 10.1093/ee/nvv002

[41] Brenes-Peralta L, Jiménez-Morales MF, Campos-Rodríguez R, De Menna F, Vittuari M. Decision-making process in the circular economy: A case study on university food waste-to-energy actions in Latin America. Energies. 2020;13:2291

[42] Keng ZX et al. Community-scale composting for food waste: A life-cycle assessment-supported case study. Journal of Cleaner Production. 2020;261:121220

[43] Awasthi SK, Sarsaiya S, Awasthi MK, Liu T, Zhao J, Kumar S, et al. Changes in global trends in food waste composting: Research challenges and opportunities. Bioresource Technology. 2020;299:122555. DOI: 10.1016/j.biotech.2019.122555

[44] Barrena R, Font X, Gabarrell X, Sanchez A. Home composting versus industrial composting: Influence of

composting system on compost quality with focus on compost stability. Waste Management. 2014;**34**:1109-1116. DOI: 10.1016/j.wasman.2014.02.008

[45] Padilla AJ, Trujillo JC. Waste disposal and households' heterogeneity: Identifying factors shaping attitudes towards source-separated recycling in Bogotá, Colombia. Waste Management. 2018;**74**:16-33

[46] ISO. 14,040 - Environmental Management - Life Cycle Assessment - Principles and Framework. Geneva, Switzerland: International Organization for Standardization; 2006

[47] Goula AM, Lazarides HN. Integrated processes can turn industrial food waste into valuable food by-products and/or ingredients: The cases of olive mill and pomegranate wastes. Journal of Food Engineering. 2015;**167**:45-50

[48] Scheel C. Beyond sustainability: Transforming industrial zero-valued residues into increasing economic returns. Journal of Cleaner Production. 2016;**131**:376-386

[49] Bhandari BR, Bansal N, Zhang M, Schuck P. Handbook of Food Powders: Processes and Properties. Amsterdam: Elsevier; 2013

[50] Karam MC, Petit J, Zimmer D, Djantou EB, Scher J. Effects of drying and grinding in production of fruit and vegetable powders: A review. Journal of Food Engineering. 2016;**188**:32-49

[51] Neacsu M, Vaughan N, Raikos V, Multari S, Duncan G, Duthie G, et al. Phytochemical profile of commercially available food plant powders: Their potential role in healthier food reformulations. Food Chemistry. 2015;**179**:159-169

[52] Bas-Bellver C, Barrera C, Betoret N, Seguí L. Turning Agri-food cooperative vegetable residues into functional powdered ingredients for the food industry. Sustainability. 2020;**12**:1284

[53] Ortiz D, Batuecas E, Orrego C, Rodríguez LJ, Camelin E, Fino D. Sustainable management of peel waste in the small-scale orange juice industries: A Colombian case study. Journal of Cleaner Production. 2020;**265**:121587

[54] Xu F, Li Y, Ge X, Yang L, Li Y. Anaerobic digestion of food waste: Challenges and opportunities. Bioresource Technology. 2018;**247**:1047-1058. DOI: 10.1016/j.biortech.2017.09.020

[55] Suhartini S, Lestari YP, Nurika I. Estimation of methane and electricity potential from canteen food waste. IOP Conference Series: Earth and Environmental Science. 2019;**230**:1-6. DOI: 10.1088/1755-1315/230/1/012075

[56] Meng Y, Li S, Yuan H, Zou D, Liu Y, Zhu B, et al. Evaluating biomethane production from anaerobic mono- and co-digestion of food waste and floatable oil (FO) skimmed from food waste. Bioresource Technology. 2015;**185**:7-13. DOI: 10.1016/j.biortech.2015.02.036

[57] Khair H, Rachman I, Matsumoto T. Analyzing household waste generation and its composition to expand the solid waste bank program in Indonesia: A case study of Medan City. Journal of Material Cycles and Waste Management. 2019;**21**:1027-1037. DOI: 10.1007/s10163-019-00840-6

[58] Soma T, Li B, Maclaren V. Food waste reduction: A test of three consumer awareness interventions. Sustainability. 2020;**12**:1-19. DOI: 10.3390/su12030907

[59] Read QD, Brown S, Cuéllar AD, Finn SM, Gephart JA, Marston LT, et al.

Assessing the environmental impacts
of halving food loss and waste along
the food supply chain. Science of the
Total Environment. 2020;**712**:1-10.
DOI: 10.1016/j.scitotenv.2019.136255

[60] Ohja S, Bußler S, Schlüter OK.
Food waste valorisation and circular
economy concepts in insect production
and processing. Waste Management.
2020;**118**:600-609. DOI: 10.1016/j.
wasman.2020.09.010

[61] KPMG. Fighting Food Waste Using
the Circular Economy. Sydney, Australia:
KPMG International; 2020. Available
from: https://assets.kpmg.com/content/
dam/kpmg/au/pdf/2019/fighting-food-
waste-using-the-circular-economy-
report.pdf

Chapter 3

Perspective Chapter: Sustainable Management and Utilization of Agricultural Waste – Innovations, Challenges, and Future Prospects

Iftikhar Ahmed, Awais Munir, Muhammad Saqlain Zaheer, Maria Ameen, Shabir Ahmad, Muhstaq Ahmad, Muhammad Iqbal and Mohamed Soliman Elshikh

Abstract

Agricultural waste (crop residues and animal manure) is a significant byproduct of farming activities. It holds potential for sustainable agriculture, renewable energy production, and environmental conservation. However, traditional disposal challenges persist, leading to air pollution and soil degradation. Innovative approaches include converting agricultural waste into bioenergy like bioethanol and biogas and producing biochar from agricultural residues, which improves soil health, water retention, and nutrient availability while sequestering carbon. Agricultural waste is also finding applications in industries. Integrating agricultural residues into industrial processes not only adds value to waste but also reduces reliance on raw materials like fossil fuels and wood. In livestock farming, crop residues are critical as feed and bedding, particularly in resource-constrained regions. While their nutritional content is often low, treatment methods can improve their digestibility. By integrating cutting-edge technologies and supportive policies, it is possible to transform agricultural waste from a burden into a resource, paving the way for a more sustainable agricultural future.

Keywords: agricultural waste, sustainable practices, renewable energy, waste management, circular economy

1. Introduction

Agricultural waste encompasses a wide range of byproducts resulting from farming activities, including crop residues, animal manure, and processing byproducts. This waste has traditionally been viewed as a disposal challenge, often leading to environmental issues such as pollution and greenhouse gas emissions. However, with the sustainable practices and technologies advancement, agricultural waste is increasingly recognized as a valuable resource. Effective management

IntechOpen

and utilization of agricultural waste are crucial for enhancing sustainability in agriculture [1]. By transforming waste into useful products such as bioenergy, biochar, and industrial materials, we can reduce environmental impacts, improve soil health, and create economic opportunities. This shift not only contributes to a more sustainable agricultural system but also aligns with broader environmental conservation goals.

Agricultural waste can be broadly categorized into three main types: crop residues, animal manure, and processing byproducts. Each type presents unique characteristics and opportunities for sustainable management and utilization [2].

Crop residues include materials left in the field after harvesting, such as maize stover, wheat straw, rice husks, corn stalks, and soybean straw [3]. These residues are rich in organic matter and nutrients, making them suitable for various applications:

- *Bioenergy production*: Crop residues can be converted into bioenergy, including bioethanol, biogas, and bio-oil. For example, maize stover and wheat straw can be processed to produce bioethanol, a renewable fuel that can replace gasoline. Similarly, anaerobic digestion of rice husks and other residues can generate biogas, which can be used for heating, electricity generation, or as a vehicle fuel.

- *Soil enhancement*: Crop residues can be returned to the soil through practices such as mulching or composting. Incorporating residues into the soil enhances soil structure, increases organic matter content, improves moisture retention, and promotes beneficial microbial activity. This not only boosts soil fertility but also helps in carbon sequestration, mitigating climate change [4].

- *Animal feed*: In resource-constrained regions, crop residues serve as a critical source of animal feed. Although their nutritional content may be lower than conventional feeds, treatments such as ensiling or ammoniation can improve their digestibility and nutritional value.

- *Industrial uses*: Certain crop residues, like rice husks, have industrial applications. Rice husks, which are high in silica, can be used to produce eco-friendly building materials, insulating boards, and bio-composites. Wheat straw is being developed into biodegradable packaging materials, reducing the reliance on plastics [5].

Animal manure, a significant byproduct of livestock farming, is rich in nutrients such as nitrogen, phosphorus, and potassium, as well as organic matter. It can be effectively managed and utilized in the following ways:

- *Composting*: Manure can be composted to produce nutrient-rich organic fertilizer. Composting stabilizes the nutrients in manure, reduces pathogens, and minimizes odors. The resulting compost can improve soil health, enhance crop yields, and reduce the need for chemical fertilizers.

- *Biogas production*: Through anaerobic digestion, animal manure can be converted into biogas and digestate. Biogas, primarily composed of methane, can be used for cooking, heating, and electricity generation. The digestate, a byproduct of biogas production, is a nutrient-rich slurry that can be used as a bio-fertilizer, enhancing soil fertility and structure.

- *Direct application*: Manure can be directly applied to fields as fertilizer. This practice recycles nutrients back into the soil, promoting plant growth. However, it requires careful management to prevent nutrient runoff and water pollution [6].

Processing byproducts are generated during the processing of agricultural products and include materials such as rice bran, molasses, fruit peels, and bagasse. These byproducts can be repurposed into valuable products:

- *Animal feed*: Many processing byproducts are rich in nutrients and can be used as animal feed. For instance, rice bran is a valuable feed ingredient for poultry and livestock due to its high protein and fat content. Molasses, a byproduct of sugar processing, is commonly used in animal feed as an energy source.

- *Bioenergy production*: Processing byproducts can also be converted into bioenergy. Bagasse, the fibrous residue from sugarcane processing, can be used to produce bioethanol or burned in boilers to generate electricity and heat. Fruit peels and other organic byproducts can be processed into biogas through anaerobic digestion.

- *Industrial and food products*: Processing byproducts can be utilized in various industrial applications. For example, fruit peels can be extracted for pectin, which is used as a gelling agent in the food industry. Bagasse is also used to produce paper, biodegradable tableware, and building materials [7].

Effective management and utilization of agricultural waste, including crop residues, animal manure, and processing byproducts, are essential for advancing sustainable agricultural practices. By transforming waste into valuable resources, we can reduce environmental impacts, enhance resource efficiency, and create economic opportunities, contributing to a more sustainable and resilient agricultural system [3].

2. Environmental impacts of agricultural waste

Agricultural waste, if not managed properly, can have significant adverse effects on the environment. This section delves into the key challenges posed by agricultural waste, such as pollution, greenhouse gas emissions, and soil degradation, followed by a case study focusing on environmental concerns in Asia due to crop residue burning (**Figure 1**).

2.1 Challenges posed by agricultural waste

- *Water pollution*: Runoff from improperly managed agricultural waste can carry nutrients, pesticides, and pathogens into water bodies. This leads to eutrophication, where excessive nutrients cause algal blooms, depleting oxygen in the water and harming aquatic life. Pathogens in the runoff can contaminate drinking water sources, posing health risks to humans and animals.

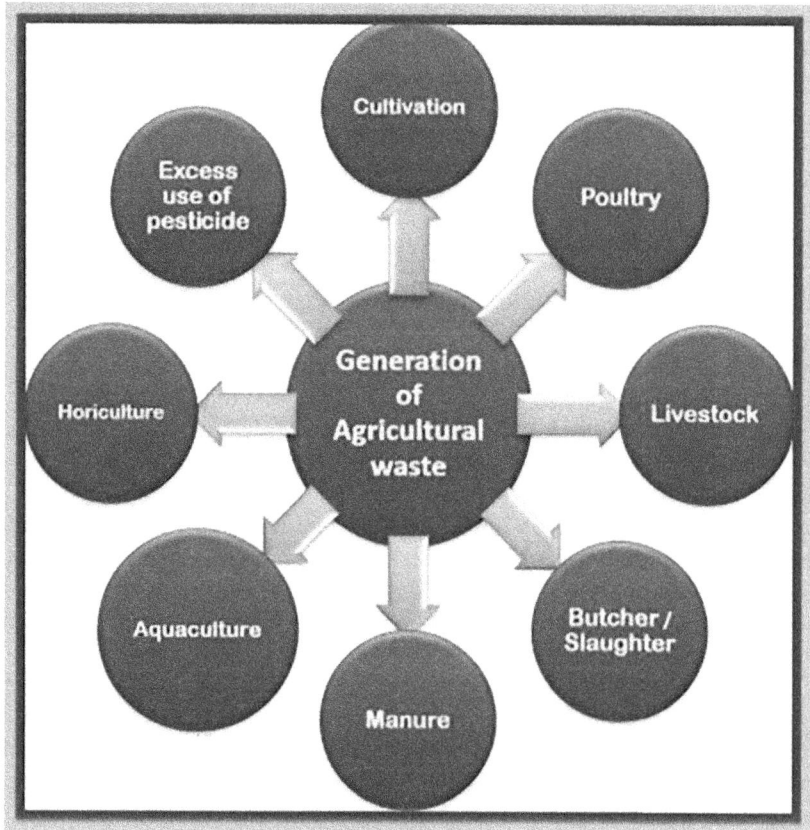

Figure 1.
Sources of generation of agricultural wastes.

- *Air pollution*: Open burning of crop residues is a common practice in many regions. This releases particulate matter, carbon monoxide, volatile organic compounds (VOCs), and other pollutants into the atmosphere, contributing to air quality deterioration. Fine particulate matter (PM2.5) is particularly harmful as it can penetrate deep into the lungs, causing respiratory and cardiovascular issues [8].

- *Methane and nitrous oxide*: Decomposition of organic agricultural waste, such as animal manure and crop residues, under anaerobic conditions produces methane (CH_4), a potent greenhouse gas. Additionally, the application of nitrogen-rich manure and fertilizers can lead to the release of nitrous oxide (N_2O), another powerful greenhouse gas. These emissions contribute significantly to global warming and climate change.

- *Carbon dioxide*: Burning agricultural waste releases carbon dioxide (CO_2) into the atmosphere. Although CO_2 from biomass burning is considered part of the carbon cycle, large-scale burning disrupts the balance and adds to the overall greenhouse gas burden.

- *Loss of soil fertility*: Continuous removal and burning of crop residues can deplete soil organic matter, essential for maintaining soil fertility and structure. Organic matter enhances soil's water-holding capacity, nutrient availability, and microbial activity.

- *Erosion*: Lack of crop residues on the soil surface leaves it exposed to wind and water erosion. Soil erosion not only reduces the land's productivity but also transports sediments and associated pollutants into waterways, further exacerbating water pollution [8].

2.2 Environmental concerns in Asia due to crop residue burning

In many Asian countries, particularly in India and China, the burning of crop residues, such as rice straw and wheat stubble, is a widespread practice. This practice is primarily driven by the need for quick and cost-effective land clearing for the next planting season [9]. However, it poses significant environmental and health concerns:

- *Severe smog and haze*: Crop residue burning during the post-harvest season leads to severe smog and haze, particularly in northern India and parts of China. The resultant poor air quality affects millions of people, leading to respiratory illnesses, eye irritation, and reduced visibility [10].

- *Health impacts*: Studies have linked high levels of particulate matter and other pollutants from crop burning to increased hospital admissions for respiratory and cardiovascular diseases. Vulnerable populations, such as children, the elderly, and those with pre-existing health conditions, are particularly at risk [10].

- *Greenhouse gas emissions*: The burning of rice and wheat residues in India alone is estimated to release millions of tons of CO_2, CH_4, and N_2O annually. These emissions significantly contribute to regional and global climate change.

- *Black carbon*: Incomplete combustion of crop residues produces black carbon, a major component of soot. Black carbon absorbs sunlight, warming the atmosphere, and when it settles on ice and snow, it accelerates melting, contributing to global warming and changes in weather patterns [11].

- *Nutrient loss*: Burning crop residues results in the loss of valuable nutrients like nitrogen, phosphorus, and potassium, which would otherwise return to the soil if the residues were decomposed naturally. This necessitates increased use of chemical fertilizers to maintain crop yields, which can further degrade soil health over time.

- *Soil structure*: The high temperatures from burning can also kill beneficial soil microorganisms and alter soil structure, making it more prone to erosion and less productive [12].

Efforts to mitigate the environmental impacts of crop residue burning in Asia include promoting alternative uses of residues, such as for bioenergy production, composting, and raw materials for industries. Additionally, policy measures and awareness campaigns are essential to encourage farmers to adopt sustainable practices. Technologies like happy seeders and zero-till farming equipment, which allow direct sowing without burning residues, are being promoted as viable solutions [13].

Overall, while agricultural waste poses significant environmental challenges, sustainable management practices can mitigate these impacts and contribute to a healthier environment and more resilient agricultural systems.

3. Innovative approaches to waste management

Agricultural waste, encompassing crop residues and animal manure, poses a significant challenge but also offers a promising opportunity for sustainable energy production. The transformation of these wastes into bioenergy sources such as bioethanol and biogas represents a forward-thinking approach to waste management, addressing both environmental and energy needs. Bioethanol, a type of biofuel, is produced through the fermentation of sugars and starches present in plant materials. This process involves breaking down complex carbohydrates into simpler sugars, which are then converted into ethanol by microorganisms like yeast. Bioethanol is a cleaner alternative to gasoline, capable of reducing greenhouse gas emissions and decreasing air pollution. Its production not only helps in managing agricultural waste but also contributes to energy diversification, reducing reliance on fossil fuels. As a renewable energy source, bioethanol supports energy security and sustainability. Similarly, biogas is another valuable bioenergy source derived from agricultural waste. Biogas is primarily composed of methane and is produced through the anaerobic digestion of organic materials, such as animal manure and crop residues. This process involves the breakdown of organic matter by microorganisms in the absence of oxygen, resulting in the production of methane-rich gas. Biogas can be used for heating, electricity generation, and as a vehicle fuel, offering a versatile and sustainable energy solution. The production of biofuels like bioethanol and biogas from agricultural waste has several environmental benefits [14]. First, it reduces the amount of waste that would otherwise contribute to pollution and land degradation. By converting waste into energy, the overall waste management system becomes more efficient and environmentally friendly. Additionally, the use of biofuels results in lower greenhouse gas emissions compared to conventional fossil fuels. This reduction in emissions is crucial for mitigating climate change and its associated impacts on the environment and human health. Furthermore, the adoption of biofuel production promotes a circular economy in the agricultural sector. Instead of viewing agricultural waste as a problem, it is seen as a valuable resource that can be utilized to generate energy [15]. This shift in perspective encourages more sustainable agricultural practices and waste management strategies. Farmers and agricultural industries can benefit economically by producing biofuels, potentially reducing their energy costs and generating additional income from the sale of bioenergy. The innovative approaches to waste management through biofuel production also have significant socio-economic implications [16]. The development of biofuel industries can create job opportunities in rural areas, supporting local economies and enhancing livelihoods. Additionally, the use of locally produced biofuels reduces the dependency on imported fossil fuels, enhancing national energy security and resilience. The transformation of agricultural waste into bioethanol and biogas exemplifies innovative waste management approaches that align with sustainability goals. These biofuels offer renewable, cleaner energy alternatives that contribute to reducing greenhouse gas emissions, promoting environmental sustainability, and supporting energy security. By adopting these approaches, the agricultural sector can play a pivotal role in addressing global energy challenges and fostering a more sustainable future [17].

4. Biochar: Transforming waste into soil enhancer

Biochar is a carbon-rich material obtained by pyrolyzing agricultural residues, such as crop husks and wood chips, at high temperatures in the absence of oxygen. This process transforms the waste into a stable form of carbon that can be applied to soil. Biochar enhances soil fertility, improves water retention, and reduces the need for chemical fertilizers, making it a valuable soil amendment [18]. Biochar contributes to soil health by enhancing nutrient retention, promoting beneficial microbial activity, and improving soil structure. Additionally, biochar acts as a carbon sink, sequestering carbon dioxide from the atmosphere and storing it in the soil for long periods. This dual benefit of improving soil quality and mitigating climate change makes biochar an innovative and sustainable approach to managing agricultural waste [19]. Rice husks and wheat straw, often considered agricultural waste, can be repurposed into eco-friendly materials and packaging solutions. These byproducts can be processed into bio-based composites, biodegradable packaging, and other sustainable products, reducing waste and the reliance on traditional, non-renewable materials. Industries can add value to otherwise discarded materials by converting agricultural waste into valuable products. This creates new revenue streams for farmers and businesses and reduces the demand for virgin raw materials. The utilization of agricultural residues in industrial applications promotes a circular economy, where waste is minimized, and resources are used more efficiently [20].

5. Industrial applications of agricultural waste

In resource-constrained regions, agricultural waste, particularly crop residues, plays a vital role in livestock farming. Crop residues, such as straw, stalks, and husks, are byproducts of harvesting cereal crops like wheat, rice, and maize. These residues serve as an essential feed resource and bedding material for livestock, addressing feed shortages and reducing the cost of animal husbandry [21]. In regions where commercial feed is either too expensive or unavailable, crop residues provide an affordable alternative. Although they are often low in protein and energy, their abundance makes them a significant component of livestock diets [22]. Ruminants, such as cattle, sheep, and goats, can utilize these residues effectively due to their unique digestive systems, which ferment fibrous plant material in their rumen. By incorporating crop residues into their diet, farmers can maintain animal health and productivity even during periods of feed scarcity [23]. Crop residues also serve as bedding material, offering a comfortable and hygienic environment for livestock (**Figure 2**). Proper bedding helps absorb moisture, reducing the risk of diseases related to poor sanitation. Moreover, bedding made from crop residues can be composted after use, turning waste into a valuable organic fertilizer that can enhance soil fertility [24]. Despite their benefits, raw crop residues often have limitations in nutritional content and digestibility. To overcome these challenges, various treatment methods can be applied to enhance their value as livestock feed [25]. Alkaline treatments using substances like sodium hydroxide or ammonia can break down the lignin in crop residues, making the cellulose more accessible to digestive enzymes. This process increases the digestibility and energy content of the residues. Ammonia treatment, in particular, also adds nitrogen, enhancing the protein content of the feed [26]. The use of microorganisms, such as fungi, to break down the fibrous components of crop residues is another

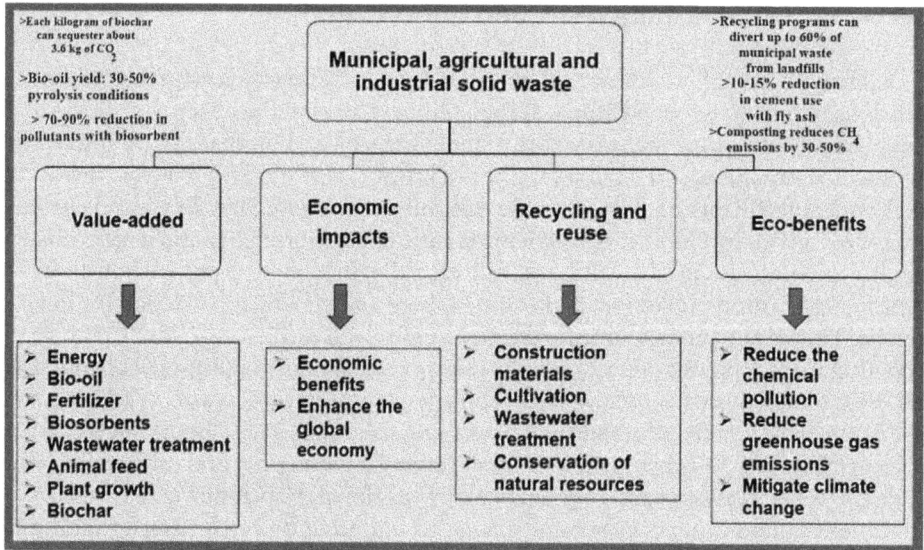

Figure 2.
Solid waste valorization: Recycling and reusing municipal, agricultural, and industrial wastes offer multiple benefits, including the creation of value-added products, environmental advantages, and economic savings.

effective method. Fungi like *Pleurotus* spp. (oyster mushroom) can degrade lignin and increase the digestibility of the residues. This process not only improves nutritional content but also adds fungal biomass, which can further enhance the protein content of the feed [27] Mechanical processing methods, such as chopping, grinding, and pelleting, can also improve the utilization of crop residues. Reducing the particle size increases the surface area for microbial action in the rumen, enhancing digestibility. Pelleting can improve the handling and storage of the feed, making it easier to transport and use [28].

6. Sustainable farming practices utilizing agricultural waste

Sustainable farming practices aim to optimize agricultural production while minimizing environmental impact. No-till farming and mulching are two techniques that effectively utilize agricultural waste to improve soil health and water retention. This method involves leaving crop residues on the field after harvest rather than plowing them under. By not disturbing the soil, no-till farming preserves soil structure, reduces erosion, and promotes the buildup of organic matter. The crop residues act as a protective layer, conserving moisture, and providing habitat for beneficial microorganisms. This practice enhances soil fertility and stability, leading to increased crop yields over time [29]. Mulching involves covering the soil with organic materials, such as crop residues, to maintain soil moisture, suppress weeds, and regulate soil temperature. Mulch decomposes over time, adding organic matter to the soil and improving its structure and nutrient content. By using agricultural waste as mulch, farmers can reduce reliance on synthetic inputs, promote sustainable soil management, and enhance the resilience of their farming systems to climate variability [30].

Integrating agricultural waste into sustainable farming techniques is a holistic approach that maximizes resource use efficiency and minimizes waste (**Figure 2**).

Agricultural waste can be composted to produce nutrient-rich organic fertilizer. Composting involves the aerobic decomposition of organic matter, converting it into humus. This process recycles nutrients, reduces the need for chemical fertilizers, and improves soil health. Composting also mitigates greenhouse gas emissions from waste decomposition by controlling the breakdown process [31]. Agricultural waste can be converted into bioenergy through processes such as anaerobic digestion and biomass gasification. Anaerobic digestion produces biogas, a renewable energy source, while the remaining digestate can be used as a fertilizer. Biomass gasification converts waste into syngas, which can be used for electricity generation or as a chemical feedstock. These technologies provide sustainable energy solutions and reduce dependency on fossil fuels [32]. Integrating trees and shrubs into farming systems, known as agroforestry, utilizes agricultural waste for mulching and soil improvement. Trees provide shade, reduce wind erosion, and enhance biodiversity, while their leaves and branches can be used as organic mulch or compost material. This practice creates a more sustainable and resilient agricultural landscape [33]. By adopting these sustainable farming practices, farmers can turn agricultural waste into valuable resources, contributing to environmental conservation, economic viability, and social well-being in rural communities.

7. Future prospects and challenges in waste utilization

The utilization of waste materials holds significant promise for sustainable development, resource conservation, and environmental protection. However, it also presents several challenges that need to be addressed through innovative technologies, strategic investments, and collaborative efforts. This discussion explores the early-stage development of waste conversion technologies, the importance of research investment, farmer awareness, and policy support, as well as the necessity for collaborative efforts among stakeholders [34].

The conversion of waste into valuable resources is a burgeoning field with numerous technological advancements emerging at the early stages of development. These technologies aim to transform agricultural, industrial, and municipal waste into bioenergy, biofertilizers, and bioproducts. For example, anaerobic digestion and composting are established methods for converting organic waste into biogas and compost, respectively. Additionally, pyrolysis and gasification processes are gaining attention for converting agricultural residues and industrial waste into biochar and syngas [35]. However, these technologies are still evolving, and several technical challenges need to be addressed. Efficiency, scalability, and cost-effectiveness are critical factors that determine the viability of these technologies. Advanced research is required to optimize the processes, improve yield, and reduce operational costs. Furthermore, there is a need for robust infrastructure to support the collection, transportation, and processing of waste materials. The development and implementation of these technologies require a multidisciplinary approach involving engineering, biotechnology, and environmental science.

Investment in research is crucial for advancing waste conversion technologies. Public and private sector funding can accelerate the development of innovative solutions and facilitate the commercialization of new technologies. Research institutions and universities play a pivotal role in conducting fundamental and applied research to address technical challenges and improve existing methods. Additionally, collaboration with industry partners can help bridge the gap between laboratory research and real-world applications. Farmer awareness and education are equally important for the successful implementation of waste utilization technologies, particularly in the agricultural sector. Farmers need to be informed about the benefits of waste management practices and trained in the use of new technologies. Extension services, workshops, and demonstration projects can play a significant role in disseminating knowledge and best practices [36]. By adopting sustainable waste management practices, farmers can enhance soil fertility, reduce reliance on chemical fertilizers, and improve crop yields, thereby contributing to environmental sustainability and economic viability. Policy support is another critical factor in promoting waste utilization. Governments can implement policies and regulations that incentivize waste management practices, such as subsidies for biofertilizers, tax breaks for companies investing in waste conversion technologies, and mandates for waste segregation at the source. Regulatory frameworks can also ensure the safe and efficient operation of waste processing facilities. Additionally, policies that promote research and development, capacity building, and infrastructure development can create an enabling environment for the growth of the waste utilization sector [37].

8. Conclusion

The successful utilization of waste requires a collaborative approach involving multiple stakeholders, including government agencies, research institutions, industry players, farmers, and non-governmental organizations. Collaborative efforts can facilitate the sharing of knowledge, resources, and expertise, thereby accelerating innovation and overcoming challenges. Public-private partnerships can play a significant role in advancing waste conversion technologies. Governments can provide funding, policy support, and regulatory oversight, while private companies can bring in technical expertise, investment, and operational capabilities. Research institutions can contribute by conducting cutting-edge research and providing scientific insights. Additionally, engaging farmers and local communities in the development and implementation of waste management practices can ensure that the solutions are practical and tailored to local conditions. International collaboration is also essential for addressing global challenges related to waste management. Countries can share best practices, technological innovations, and policy frameworks to enhance their waste utilization efforts. Global initiatives and forums can facilitate the exchange of knowledge and experiences, thereby fostering a collective approach to waste management. Despite the promising prospects, several challenges remain. Technical limitations, high initial costs, lack of infrastructure, and resistance to change are some of the obstacles that need to be addressed. Moreover, ensuring the environmental and social sustainability of waste conversion processes is crucial. Life cycle assessments, environmental impact assessments, and stakeholder consultations can help in identifying and mitigating potential negative impacts. In conclusion, the future prospects of waste utilization are bright, with significant potential for sustainable development and environmental protection. However, realizing this potential requires early-stage technological advancements, substantial research investments, farmer awareness, policy support, and collaborative efforts among stakeholders. By addressing these challenges and leveraging innovation, we can pave the way for a sustainable and resource-efficient future.

Perspective Chapter: Sustainable Management and Utilization of Agricultural Waste – Innovations...
DOI: http://dx.doi.org/10.5772/intechopen.1009015

Acknowledgements

The author acknowledges the use of AI tools for language polishing of the manuscript.

Conflict of interest

The authors declare no conflict of interest.

Author details

Iftikhar Ahmed[1], Awais Munir[2], Muhammad Saqlain Zaheer[1],
Maria Ameen[3,4]*, Shabir Ahmad[5], Muhstaq Ahmad[4], Muhammad Iqbal[3]
and Mohamed Soliman Elshikh[6]

1 Department of Agricultural Engineering, Khawaja Fareed University of Engineering
and Information Technology, Rahim Yar Khan, Pakistan

2 Institute of Agro-Industry and Environment, The Islamia University of Bahawalpur,
Pakistan

3 Department of Botany, University of Chakwal, Chakwal, Punjab, Pakistan

4 Plant Systematics and Biodiversity Lab, Department of Plant Sciences,
Quaid-i-Azam University, Islamabad, Pakistan

5 Northwest Institute of Eco-environment and Resources, Chinese Academy of
Sciences (CAS), Lanzhou, China

6 Department of Botany and Microbiology, College of Science, King Saud University,
Riyadh, Saudi Arabia

*Address all correspondence to: mameen@bs.qau.edu.pk

IntechOpen

References

[1] Koul B, Yakoob M, Shah MP. Agricultural waste management strategies for environmental sustainability. Environmental Research. 2022;**206**:112285

[2] Sharma P et al. Sustainable organic waste management and future directions for environmental protection and techno-economic perspectives. Current Pollution Reports. 2024;**10**(3):1-19

[3] Rathour RK et al. Recent trends, opportunities and challenges in sustainable management of rice straw waste biomass for green biorefinery. Energies. 2023;**16**(3):1429

[4] Chaudhary B, Kumar V. Emerging technological frameworks for the sustainable agriculture and environmental management. Sustainable Horizons. 2022;**3**:100026

[5] Duque-Acevedo M et al. Agricultural waste: Review of the evolution, approaches and perspectives on alternative uses. Global Ecology and Conservation. 2020;**22**:e00902

[6] Sarangi PK et al. Recent progress and future perspectives for zero agriculture waste technologies: Pineapple waste as a case study. Sustainability. 2023;**15**(4):3575

[7] Hernandez D et al. The role of artificial intelligence in sustainable agriculture and waste management: Towards a green future. International Transactions on Artificial Intelligence. 2024;**2**(2):150-157

[8] Ouyang H et al. Agricultural waste-derived (nano) materials for water and wastewater treatment: Current challenges and future perspectives. Journal of Cleaner Production. 2023;**421**:138524

[9] Ameen M et al. Assessing the bioenergy potential of novel non-edible biomass resources via ultrastructural analysis of seed sculpturing using microscopic imaging visualization. Agronomy. 2023;**13**(3):735

[10] Blasi A et al. Lignocellulosic agricultural waste valorization to obtain valuable products: An overview. Recycling. 2023;**8**(4):61

[11] Kari ZA et al. Recent advances, challenges, opportunities, product development and sustainability of main agricultural wastes for the aquaculture feed industry–a review. Annals of Animal Science. 2023;**23**(1):25-38

[12] Phiri R, Rangappa SM, Siengchin S. Agro-waste for renewable and sustainable green production: A review. Journal of Cleaner Production. 2023;**434**:139989

[13] Jabeen S et al. Ultra-sculpturing of seed morphotypes in selected species of genus salvia L. and their taxonomic significance. Plant Biology. 2023;**25**(1):96-106

[14] Ameen M et al. Hazardous waste management of novel non-edible Platycladus orientalis seed oil via recyclable yttria-based phyto-nanocatalyst: A practical approach towards sustainable bioenergy conversion. Sustainable Energy Technologies and Assessments. 2024;**67**:103845

[15] Ameen M et al. Wild melon: A novel non-edible feedstock for bioenergy. Petroleum Science. 2018;**15**:405-411

[16] Aziz A et al. Microscopic techniques for characterization and

authentication of oil-yielding seeds. Microscopy Research and Technique. 2022;**85**(3):900-916

[17] Khedulkar AP et al. Sustainable high-energy supercapacitors: Metal oxide-agricultural waste biochar composites paving the way for a greener future. Journal of Energy Storage. 2024;**77**:109723

[18] Adisa O et al. A comprehensive review of redefining agricultural economics for sustainable development: Overcoming challenges and seizing opportunities in a changing world. World Journal Of Advanced Research and Reviews. 2024;**21**(1):2329-2341

[19] Kumar JA et al. Agricultural waste biomass for sustainable bioenergy production: Feedstock, characterization and pre-treatment methodologies. Chemosphere. 2023;**331**:138680

[20] Mujtaba M et al. Lignocellulosic biomass from agricultural waste to the circular economy: A review with focus on biofuels, biocomposites and bioplastics. Journal of Cleaner Production. 2023;**402**:136815

[21] Bumharter C et al. New opportunities for the European biogas industry: A review on current installation development, production potentials and yield improvements for manure and agricultural waste mixtures. Journal of Cleaner Production. 2023;**388**:135867

[22] Ameen et al. Prospects of Bioenergy Development in Future, Encyclopedia of Renewable Energy, Sustainability and the Environment (First Edition), Elsevier, 2024. pp. 497-508. ISBN 9780323939416. Available from: https://www.sciencedirect.com/science/article/pii/B9780323939409000244. DOI: 10.1016/B978-0-323-93940-9.00024-4

[23] Sangmesh B et al. Development of sustainable alternative materials for the construction of green buildings using agricultural residues: A review. Construction and Building Materials. 2023;**368**:130457

[24] Zaib M et al. A review on challenges and opportunities of fertilizer use efficiency and their role in sustainable agriculture with future prospects and recommendations. Current Research in Agriculture and Farming. 2023;**4**(4):1-14

[25] Van Tran T et al. A critical review on pineapple (Ananas comosus) wastes for water treatment, challenges and future prospects towards circular economy. Science of the Total Environment. 2023;**856**:158817

[26] Jaffur BN et al. Current advances and emerging trends in sustainable polyhydroxyalkanoate modification from organic waste streams for material applications. International Journal of Biological Macromolecules. 2023;**253**:126781

[27] Ameen M et al. Biodiesel synthesis from Cucumis melo var. Agrestis seed oil: Toward non-food biomass biorefineries. Frontiers in Energy Research. 2022;**10**:830845

[28] Arias A, Feijoo G, Moreira MT. Biorefineries as a driver for sustainability: Key aspects, actual development and future prospects. Journal of Cleaner Production. 2023;**418**:137925

[29] Ameen M et al. Conversion of novel non-edible Bischofia javanica seed oil into methyl ester via recyclable zirconia-based phyto-nanocatalyst: A circular bioeconomy approach for eco-sustenance. Environmental Technology & Innovation. 2023;**30**:103101

[30] Xu P et al. Pretreatment and composting technology of agricultural organic waste for sustainable agricultural development. Heliyon. 2023;**9**(5):e16765

[31] Garg R, Sabouni R, Ahmadipour M. From waste to fuel: Challenging aspects in sustainable biodiesel production from lignocellulosic biomass feedstocks and role of metal organic framework as innovative heterogeneous catalysts. Industrial Crops and Products. 2023;**206**:117554

[32] Varghese SA et al. Renovation of agro-waste for sustainable food packaging: A review. Polymers. 2023;**15**(3):648

[33] Roy S, Rautela R, Kumar S. Towards a sustainable future: Nexus between the sustainable development goals and waste management in the built environment. Journal of Cleaner Production. 2023;**415**:137865

[34] Ameen M et al. Prospects of catalysis for process sustainability of eco-green biodiesel synthesis via transesterification: A state-of-the-art review. Sustainability. 2022;**14**(12):7032

[35] Ameen M et al. Cleaner biofuel production via process parametric optimization of nonedible feedstock in a membrane reactor using a titania-based heterogeneous nanocatalyst: An aid to sustainable energy development. Membranes. 2023;**13**(12):889

[36] Mwantimwa K, Ndege N. Transferring knowledge and innovations through village knowledge center in Tanzania: Approaches, impact and impediments. VINE Journal of Information and Knowledge Management Systems. 2024;**54**(2):379-397

[37] Rozina et al. Implication of scanning electron microscopy as a tool for identification of novel, nonedible oil seeds for biodiesel production. Microscopy Research and Technique. 2022;**85**(5):1671-1684

Environmental Consequences of Crop Residue Burning and Its Threats to Urbanization

Prateek Singh and Jay Shankar Singh

Abstract

Crop residue burning and urban sustainability threats are vastly interrelated themes that have predominant impacts on environmental quality, public health, social justice, and resource fulfillment. Over recent years, CR burning has become a common practice in developing countries. Due to the lack of a proper disposal mechanism for CR, the farmer burns them at agricultural farm sites and releases greenhouse gases into the atmosphere, which poses a great threat to the environment. Burning crop residue is a severe ecological issue that impacts both rural and urban regions substantially. Crop residue burning is a ubiquitous agricultural practice in numerous parts of the world, especially in developing nations, used to clear fields rapidly and manage the crop residues. These practices will operate in the short term, which has substantial long-term impacts on atmospheric conditions, soil health, and global (CH_4) dynamics. Furthermore, the present study evaluates and elaborates on the environmental consequences of crop residue burning and its threats to urbanization.

Keywords: crop residue burning, air pollution, urbanization challenges, greenhouse gas, environmental impact

1. Introduction

The rapid urbanization and expansion of the population are continuously putting a burden on agronomy to provide food to fulfill human needs and other living creatures [1]. Urbanization affects the agricultural practices that, in turn, also affect the sustainability of the environment. One of the major threats to challenge the rapid urbanization is the management of agricultural trash, especially crop residue burning. Nowadays, farmer burning of crop waste is currently one of the biggest problems facing agriculture, especially in regions like Southeast Asia and the Southern region, where rice and wheat are the most prevalent cropping patterns. The negative impacts of these actions on the environment, which have a direct effect on human well-being, have grown more visible in urbanized areas, such as cities, and continue to develop and encroach on natural places [2]. The long-term effects of this practice are most severe in urban dwellers who suffer from elevated air pollution and related health issues because of the dependency of rural communities on crop residue burning (CRB) as a short-term solution. The widespread agricultural practice of burning

IntechOpen

crop residues to prepare fields for the next harvesting season has serious detrimental effects on public health and the environment, posing a significant threat to urban habitats. Farmers usually resort to burning agricultural trash since other methods of disposing of it can be expensive and time-consuming [3]. In order to control the problem and issue, we must solve it; technology innovators and urban stakeholders must be involved.

2. Global and regional perspectives on CRB

Crop residue burning is an important global problem that threatens the environment and human health in cities. Agricultural activities, such as crop residue burning, generate between 10 and 12% of total anthropogenic greenhouse gas emissions, which renders the agriculture sector a major contributor to global climate change [4]. The problems with CRB are exacerbated by urbanization, particularly as growing cities encroach on agricultural lands. Urban air quality is greatly affected by the emission of particulate matter (PM2.5 and PM10), carbon monoxide, and volatile organic compounds during CRB, which worsens the pollution that is already produced by industry and automobiles. In regions like North India, for example, CRB in rural areas has a direct and severe effect on surrounding metropolitan centers like Delhi, which leads to seasonal smog epidemics and serious health consequences [5]. In a similar vein, Southeast Asia's "haze season," triggered by crop residue burning, affects large cities like Jakarta and Bangkok, as well as distant urban areas in nearby countries. Despite the detrimental ecological consequences, the desire to clear fields for the next planting season frequently leads to the sustained practice of burning agricultural leftovers. The continued expansion of urban areas diminishes accessible agricultural land and diverts policy focus from rural problems, inhibiting the shift to sustainable residue management techniques. In spite of that, urbanization also creates opportunities for innovative solutions. Cities can act as a center point for promoting sustainable technologies like biochar production, composting, and waste-to-energy-generating systems that utilize crop residues. With collaborative policies that integrate urban and rural environmental strategies will be essential to address CRB's dual impact on agriculture and urban living. In this interconnected landscape, tackling CRB demands a coordinated global and regional approach, balancing urban development with sustainable agricultural practices to mitigate health and environmental risks [6]. Crop residue burning produces water contamination in both surface and underground water sources that negatively impacts urban water supply systems. When smoke particles escape the atmosphere during burning, they accumulate on urban surfaces before rain brings them into drainage systems where they cause water quality deterioration through pipe blockages. The runoff contains dangerous chemicals and small particles that filter into groundwater sources, which produces groundwater contamination. Municipal water treatment operations become more expensive while achieving reduced efficiency because of pollutants that exist in water supplies. CRB affects rainwater harvesting because the deposition of soot on collected water makes it unusable for household and agricultural activities. Pollutant deposition in urban wetlands creates ecological damages while threatening the biodiversity of the area. The urgent requirement for whole-water resource management systems should be adopted in urban development master plans for places exposed to seasonally devastating crop residue burning incidents.

3. Impact on urbanization threats due to CRB

3.1 Urban air quality degradation due to CRB

The growth of cities typically leads to a substantial deterioration of air quality due to the influx of industry, automobiles, and energy consumption in cities (**Table 1**) [8]. The presence of standard air quality gives rise to serious environmental and public health issues, especially within the emergence of metropolitan areas where infrastructure strives to keep down. Urbanization and the clustering of people in cities has been one of the most significant shifts in human history [9]. While it stimulates prosperity, technical innovation, and greater access to services, it often comes at the expense of harmful emissions, in particular in terms of air quality. The incineration of agricultural waste plays a major role in deteriorating urban air quality, which is mainly dependent on farmland or nearby cities [10]. This activity emits a significant quantity of particulate matter (PM2.5 and PM10), volatile organic compounds (VOCs), carbon monoxide, and black carbon into the air. These pollutants generate dense smog layers that distort the skyline, impair visibility, and worsen the air quality to hazardous levels, which travel through the wind into nearby towns and cities [11]. In addition to disrupting the activities of day-to-day life, the visibility issue diminished the allure of urban environments, which reduced the visibility issue and concealed iconic sites and natural views that interfered with daily activities. Heightened safety concerns associated with visibility cause more road accidents and interruptions to air travel and hazy clouds, restricting outdoor activities like cycling, running, and leisure, which had a major negative impact on city dwellers' physical and mental health. The decline in air quality further intensifies public health issues, especially for children, the elderly, and those with existing respiratory or cardiovascular ailments [12]. The burning crop trash was generating a fine particulate sediment that enters the bloodstream and lungs to increase the risk of respiratory diseases like asthma attacks, which potentially is a serious long-term problem, including lung cancer. Additionally, the social and economic ramifications are significant, characterized by escalating healthcare expenses and diminished productivity stemming from health-related issues. Furthermore, the social and economic impact is substantial to urbanization.

4. Impact of CRB on urbanization

Urbanization has profoundly changed the Farmland Ecological Dynamics and Land Use Change Pattern by frequent encroachment onto the farmland [13]. Burning crop trash is widely practiced as a method of disposal, air quality, and its impact on soil health. The smoke and particulate matter produced during this process can travel far, degrading air quality in nearby urban areas. Not only increases harmful emissions but also poses serious health risks, mainly respiratory and cardiovascular problems. The resulting pollution can disrupt daily life, drive up healthcare costs, and put additional pressure on urban infrastructure. The locations of agricultural areas in relation to urban areas induce the problem even more, constituting a bypass for pollutants to infiltrate the settled areas. A detailed plan of action that utilizes the combination of stringent legal measures and good holistic management of agricultural wastes must be developed to deal with the problem adequately [14]. This approach attempts to abate the adverse effects of burning crop residues on public health and urban ecosystems.

Aspect	Impact	Source Pollutants	Health Implications	Sources	References
Volatile Organic Compounds	Release of VOCs during CRB	PM_{10}, $PM_{2.5}$	Climate change has an indirect impact on health, such as heatwaves and vector-borne illnesses.	The combustion of agricultural leftovers releases stored carbon.	[1, 2]
Greenhouse Gases (GHGs)	Elevated amounts of CO_2, CH_4, and N_2O trigger global warming.	Toluene, Benzene, Formaldehyde	Continuous exposure is connected with respiratory tract and eye inflammation and cancer.	Pyrolysis of organic material in residues	[3]
Toxic Gases	Smog is generated when $PM_{2.5}$ and PM_{10} levels rise.	PM_{10} $PM_{2.5}$	Cardiovascular disorders can cause respiratory issues, reducing lung function and worsening asthma symptoms.	Burning of wheat residue, paddy stubble.	[4, 5]
Urban Heat Island Effect	Increase in localized temperatures due to atmospheric pollutants trapping heat	Greenhouse gas PM.	Enhanced energy use for cooling and heat stress in vulnerable populations	Accumulation of black carbon and greenhouse gases	[6]
Chemical Transformation	Chemical interactions between VOCs and NO_x lead to the rise of secondary pollutants such as ozone.	Fly ash, Black carbon,	Alters esthetic the visual appeal of municipal infrastructure which effect on photosynthesis in plants.	soot and Windborne ash from burning fields	[7]
Deposition of Soot and Ash	Deposition of soot on plants, buildings, and other surfaces	SO_2, CO, NO_x,	Symptoms of asthma and bronchitis worsen.	Photochemical reactions in the urban environment	[8]
Airborne Particulate Matter	Increased amounts of hazardous gasses.	Ground level ozone	Symptoms of asthma and bronchitis worsen.	Photochemical reactions in the urban environment	[9]

Table 1.
Impact of crop residue burning (CRB) on human health and air quality.

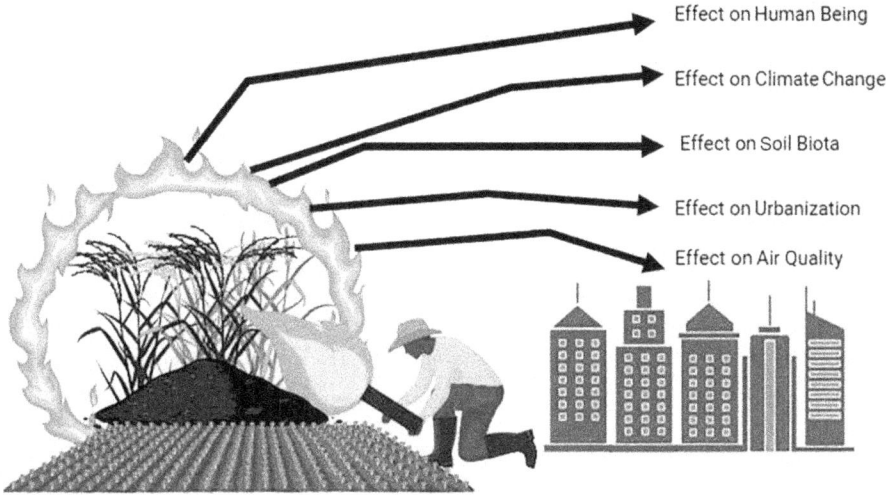

Figure 1.
Effects of CRB on urbanization.

The method of burning crop residues, paddy straw, for instance, is a significant threat to urbanization due to its extensive negative effects on the environment and human health [15]. Such burning releases a variety of harmful pollutants, including carbon monoxide, black carbon, fine particulate matter (PM2.5 and PM10), and greenhouse gases. Air quality in urban centers is severely polluted due to the immense distances covered by these pollutants and the extensive use of the phrase. The worsening pollution in cities caused by industrial and automobile emissions leads to extreme smog and comparatively low visibility. Urban life faces numerous challenges, including heightened risks of respiratory and cardiovascular illnesses, rising healthcare costs, and strain on medical facilities [16]. Urban expansion will be profoundly influenced by the financial challenges associated with mitigating these adverse effects, as well as by productivity declines stemming from inadequate public health and interrupted transportation systems. Additionally, the disruption of regional weather patterns by climate-related changes may exacerbate urban water shortages, leading to the formation of heat islands and decreased rainfall [17]. And reliance on the burning of crop residues will underscore the lack of sustainable agricultural practices, thereby emphasizing the necessity for integrated strategies to manage the interactions between urban and rural areas and to protect the cities from the associated detrimental impacts (**Figure 1**).

5. Impact of CRB on urban water systems

Urban water systems are severely affected by crop residue burning (CRB), which presents serious problems to controlling and sustaining water quality in urban areas (**Table 2**) [26]. Particulate matter and hazardous chemicals emitted during CRB come up in urban water sources such as lakes, reservoirs, and rivers, which are essential sources of drinking water. These pollutants, which include heavy metals, numerous chemical compounds, and fine particulate matter (PM2.5 and PM10), change the pH of water, enhance turbidity, and contaminate hazardous substances in water [27]. Urban rivers and lakes, which are already under pressure from industrial discharges

Environmental consequences	Effects	References
Eutrophication	Eutrophication is the excessive enrichment of water bodies with nutrients.	[18]
Stormwater Clogging	making water unsafe and unpleasant, Harmful chemicals leach into aquifers.	[19]
Water Treatment Challenges	Increased pollutants strain treatment systems and raise costs.	[20]
Reduced Groundwater Recharge	Infiltration rates are decreased by land deterioration.	[21]
Wetland Degradation	Pollutants disrupt urban wetland ecosystems and biodiversity.	[22]
Rainwater Harvesting Issues	Soot and ash contaminate harvested rainwater	[23]
Surface Water Pollution	Ash and soot settle in water bodies, reducing water quality	[24]
Groundwater Pollution.	Ash clogs drains and enhances urban flood threats.	[25]

Table 2.
Impact of crop residue burning (CRB) on urban water systems and management.

and municipal waste, are further impacted by CRB fallout, exacerbating the difficulties in ensuring safe water for city residents.

Urban water treatment facilities are greatly influenced by the rise in contamination levels, and the presence of pollutants associated with CRB necessitates enhanced filtration and chemical treatment processes to comply with safety regulations, resulting in increased operational expenses and reduced efficiency [28]. The act of burning crop leftovers greatly affects road usage through air pollution and lowering visibility. Together with atmospheric conditions, haze from burning trash makes the environment extremely dangerous, especially during deep fog in the early mornings and evenings. The increase of accidents caused by misty conditions differs from minor ones to the more serious crashes, causing danger to the passengers and other people using the road. Lack of safety on the roads reduces daily commute efficiency. These public transportation systems, such as trains and busses, are greatly impacted as they have to operate at a much slower speed to prevent accidents. Commuters were frustrated as it reduced service reliability and slowed travel times. Additionally, respiratory issues became a lot harder to manage due to air containing fine particles (PM10 and PM2.5), making traveling in vehicles that were either not fully enclosed or properly ventilated incredibly difficult. Not only does this interfere with everyday movement, but it also affects the quality of life in growing cities. CRB affects the air travel system and has adverse effects on urban connection and integration [29].

This situation leads to a higher frequency of maintenance and repair requirements, placing additional pressure on municipal finances. The escalating expenses associated with addressing water contamination tied to CRB frequently compel cities to reallocate funds from other essential urban development sectors, including housing and public transportation [30]. These disruptions in water distribution networks caused by sedimentation related to CRB create disparities in water access among urban communities, worsening social inequalities as cities expand; the need for clean water increases, making the effects of CRB on water systems a critical challenge associated with urbanization, and tackling this problem necessitates a collaborative effort to lower CRB emissions and enhance urban water infrastructure resilience. But, by implementing advanced water treatment solutions, launching public education initiatives on water conservation, and enforcing stricter regulations on CRB activities, we may help in protecting urban water quality and maintaining the sustainability of water systems amid any environmental pressures.

6. Impact of CRB on urban transportation

Crop residue burnings seriously reduce air quality while lowering roadway visibility to dangerous levels especially during foggy conditions of early morning and evening [5]. The atmospheric effects that combine during commuting periods result in dangerous road conditions that lead to varied car collision risks including minor and major accidents. The situation results in reduced road safety standards and diminished travel efficiency on a daily basis. Multiple agencies maintain control over public transportation systems because they need to maintain reduced speeds to protect passengers from safety hazards, which creates delays and impacts public service dependability [31]. Fine particulate matter PM10 and PM2.5 creates worsened respiratory conditions and disrupts vehicular ventilation systems primarily in locations with congested roadways. The peak CRB time period triggers frequent operational disruptions along with health issues across the entire air transportation network. Maintenance disruptions resulting from CRBs cause problems for passengers while allowing damage to business operations and tourism sectors which diminishes economic connections and productivity levels [32]. The combined implications in urban centers put strain on municipal systems, exposing the fundamental vulnerabilities of expanding metropolises [33]. Resolving CRB impacts on transportation necessitates measures to improve urban flexibility while also implementing sustainable mobility alternatives. It is common for CRB seasons to cause flight delays, cancelations, and operational inefficiencies at urban airports, which are often a lifeline for both business and tourism.

In addition to inconveniencing passengers, this also has cascading effects on local economies that rely on the timely transportation of goods and people [32]. Industries in urban areas reliant on air flight, such as e-commerce and perishables, are particularly hard hit, amplifying economic vulnerabilities [33]. Prolonged exposure to polluted air in urban airports strains public health infrastructure and reduces worker productivity. Urbanization and CRB have intertwined challenges associated with disruptions in urban transportation and infrastructure and have a greater impact on urban communities as population increases and mobility demands increase. Integrated approaches that address CRB emissions and enhance the resilience of urban infrastructure are necessary to address these issues and ensure that cities continue to thrive as social and economic centers despite the environmental challenges.

7. Impact of CRB on climate change and urban resilience

Urban pollution caused by metropolitan cities is enhanced by crop residue burning (CRB). This is particularly true in regions wherein agricultural fields are adjacent to urban centers. One of the most significant impacts of CRB is its role in elevating PM2.5 and PM10 levels, which directly leads to respiration issues and chronic diseases [5]. CRB indeed releases extremely high amounts of PM2.5 and PM10 residues, which are then carried over long distances, eventually making their way to urban areas. Combustion of agricultural remnants releases both PM10 and PM2.5, which severely worsens the air quality of urban areas, resulting in it being hazardous [34]. During this smog-exacerbated season, a group of harmful pollutants drastically deteriorates the air environment. Even when these particular types of particulate matter are slightly larger, they still have the potential of lowering visibility, thus harming eyes, nose, and throat [35]. To draw an estimate, the sheer amount of smog witnessed within agricultural areas is because of PM emissions due to CRB.

8. Urbanization's impact on soil microflora due to CRB

The impact on soil microflora, which is essential for maintaining soil health and ecosystem services, is significant due to urbanization and crop residue burning [36]. Soil sealing, compaction, and contamination due to urbanization can result in decreased habitat availability and diversity for soil microorganisms. The disruption of microbial communities is a result of these changes, which affect processes like nutrient cycling, organic matter decomposition, and soil structure maintenance [37]. Burning residues from crops renders these problems harsher by slaying soil microflora directly with severe heat and releasing particulate matter and potentially hazardous pollutants like polycyclic aromatic hydrocarbons (PAHs) into the soil and atmosphere. The alteration of soil pH, reduction of organic carbon, and depletion of microbial biomass caused by this practice result in a decline in microbial diversity. Additionally, the long-term soil degradation caused by urbanization and residue burning is because of the synergistic effects of urbanization and residue burning, where urbanization and residue burning are threats to agricultural productivity and ecological stability. The restoration of soil microbial health and mitigation of these impacts requires sustainable agricultural practices, such as residue management and urban green planning.

9. Urban productivity losses due to CRB

The environmental and social consequences of agricultural practices, such as crop residue burning, are deeply interconnected with the economic burden of managing air quality in urban areas. Crop residue burning is a very common practice in agricultural regions and contributes to the emission of high levels of particulate matter (PM2.5 and PM10), greenhouse gases, and air pollutants into the atmosphere [5]. These compounds are then transported to the closest cities, worsening pollution and making it extremely difficult to manage air quality in those areas. Cities are required to invest in a variety of measures in order to mitigate the negative effects of this seasonal influx of pollutants, including air filtration systems, air monitoring stations, and campaigns to promote health and well-being. The increased burden on public health systems due to respiratory diseases, heart diseases, and other pollution-related diseases has made the situation worse. Major damage is also caused to cities' productivity due to this practice. As pollution increases, there are increased rates of nonappearance and lower productivity in both indoor and outdoor settings. People working in outdoor industries like construction and transportation are at a higher risk of exposure to polluted air, which contributes to lower productivity levels. On the other hand, in the case of businesses, this creates a negative impact due to higher medical costs, loss of branding, and low levels of customer participation since people try to avoid going outside.

10. Urban threats and its communities

The practice of burning crop residue (CRB) worsens the social and cultural injustices in urban settings, especially among the disadvantaged sections of the community, particularly the low-income urban inhabitants such as slum dwellers. These communities are often deprived of basic healthcare amenities, air filters, and clean water facilities, making them most vulnerable to the deleterious effects of the pollution stemming from CRB [38]. The lack of adequate resources and poor site conditions exacerbate the situation, and hence, these socioeconomic groups face enormous challenges pertaining to air and water quality. The unequal consequences of CRB have led to social unrest and protests in urban centers where citizens are advocating for elementary rulemaking regarding CRB activities and the urban planners to incorporate other useful measures to avoid these adverse effects. This civil agitation is the manifestation of what is increasingly felt to be a political soup whereby social inequalities are embedded as pollution struggles, such as these urban inequities, are just the tip of the iceberg. Pollutants generated by CRB and the overall urban-crime economic lifestyle are also serious impediments to urban civilization and well-being. Episodes of high pollution levels make the general populace constrained to their indoors, and this greatly reduces outdoor social interaction and attendance of recreational facilities, including parks, playgrounds, and promenades. Also restricted are most public gatherings, such as events, festivals, and cultural occasions. Restrictions imposed on outdoor activities reduce community participation and negatively impact both physical and mental well-being since there is limited social contact and opportunities to access spaces for exercise and relaxation. CRB has caused a drastic decline in cultural and leisure activities in cities, which include tourism, sports, and even urban festivals [39]. Tourists tend to avoid visiting cities with increasing levels of air pollution, which leads to a decline in the hospitality and tourism sectors. Due to the rise in air pollution, many outdoor urban sporting events, marathons, and competitions around the city are either canceled or lack participation, interrupting the city's cultural calendar. Similarly, in the case of urban sporting events, a pollution crisis takes precedence over the identity and heritage being celebrated by the festivals, subsequently having a negative impact on the people and the economy in that area. The social and cultural conflicts that arise as a result of CRB exemplify how climate change endangers the growth and sustainability of cities [40]. Addressing these problems will help improve the quality of life in urban settings and encourage social justice. Targeted measures such as improving health access for excluded communities and establishing recreational facilities that are immune from pollution are needed to address these negative impacts and safeguarding urban communities.

11. Mitigation strategies of urbanization and CRB impact

Restrictions imposed on outdoor activities reduce community participation and negatively impact both physical and mental well-being since there is limited social contact and opportunities to access spaces for exercise and relaxation [41]. CRB has caused a drastic decline in cultural and leisure activities in cities, which include tourism, sports, and even urban festivals. Tourists tend to avoid visiting cities with increasing levels of air pollution, which leads to a decline in the hospitality and tourism sectors. Due to the rise in air pollution, many outdoor urban sporting events, marathons, and competitions around the city are either canceled or lack participation, interrupting the city's cultural calendar [42]. Similarly, in the case of urban sporting events, a pollution crisis takes precedence over the identity and heritage being celebrated by the festivals, subsequently having a negative impact on the people and the economy in that area. The social and cultural conflicts that arise because of the CRB exemplify how climate change endangers the growth and sustainability of cities [40]. Addressing these problems will help improve the quality of life in urban settings and encourage social justice. Targeted measures such as improving health access for excluded communities and establishing recreational facilities that are immune from pollution are needed to address these negatives [43]. Simultaneously, the focus on developing alternative agricultural practices, such as out-in-the-field crop residue processing machinery and technologies for converting these residues into biofuels, can minimize the dependence on burning. It is also possible for urban innovation centers to partner with rural areas and jointly create scalable technologies that meet agricultural and urban demands within the scope of the Urban-Sustainability Innovation Alliance (USIA). USIA is a collaborative framework that aims to connect urban technology innovation with rural agricultural sustainability, allowing for combined solutions to difficulties such as pollution control, crop residue management and infrastructure development. Community and urban involvement are crucial for the success of integrated considerations within the framework of CRB solution efforts. Urban dwellers can engage in strategies at the community level when they get involved in advocacy programs promoting clean air and different air quality management policy initiatives [44]. The actions of urban and rural communities need to be linked because CRB, which is largely done in rural areas, has effects that affect urban centres. Proposals like paying farmers, subsidizing the purchase of crop residue management machinery, and rural programs on the economic advantages of looking after the soil adequately would help reduce such polarization. Locally initiated strategies such as tree planting, installation of air filters, or monitoring by citizens may enhance policy initiatives and technological approaches in which people actively take responsibility for not doing such things. By all these means, urbanization being the process that it is, we may be able to cope with some of the challenges posed by CRB, for example, air pollution, reducing this way.

12. Residue burning falling and its implication

Combining satellite monitoring of activities and AI-based modeling approaches can potentially change the way we manage crop residue burning. Satellites mounted on devices can supply information in real time with regard to burning events, enabling the characterization of intakes and trends within the agricultural environment [45]. AI can utilize these datasets to track burning patterns and determine its implications on environmental conditions, thus enabling timely action by government leaders. Information can also be made available through mobile applications and other online platforms in order to raise public awareness of effective alternative methods and promote their adoption. Moreover, the combination of remote sensing with ground-based sensors can enhance the performance of air quality systems and elicit preventive responses. This work should focus on broadening the range of countries that would be able to employ these technologies in an uncomplicated manner.

13. Sustainable urbanization and its approaches

This is the current policy, which the interrelated problem of crop residue burning and increased urbanization should be addressed by all levels of governments, industries, universities, and communities. Governments should be put in place and ensure that there are policies aimed at promoting reductions in burning while incentivizing farmers' management practices for residue waste [46]. Industries can support efforts by venturing into the production of energy from residue material and constructing energy-efficient building materials from agricultural waste. NGOs and educational institutions play a vital role in teaching farmers about sustainable farming methods while providing them training. In order to ensure that cities and peri-urban areas have the capacity to cope with the trash created by surrounding rural regions, urban planners ought to integrate agricultural waste management into wider environmental sustainability frameworks. A more sustainable and inclusive approach to addressing these sorts of problems may be developed by establishing public-private partnerships and promoting community-driven enterprises.

14. Summary and conclusion

Crop residue burning is a ubiquitous issue with substantial environmental, agricultural, and urban effects. The deliberate release of greenhouse gases (GHGs) such as carbon dioxide, methane, and nitrous oxide contributes to climate change, whereas particulate matter and other pollutants ruin air quality, which threatens human health. In fields, residue burning reduces soil fertility by removing beneficial microbes, which eliminates organic matter decomposition. Urban areas are impacted through transboundary pollution, which increases pressure on straining infrastructure and healthcare systems. The findings are highlighted in the multifaceted approach to address these challenges. Effective policies, such as enforcement of anti-burning regulations for the provision of subsidies for alternative residue management practices, are necessary. Innovations in technologies like satellite monitoring are driven by AI models, which may boost the identification and mitigation of fire events. Furthermore, education campaigns and communities are involved in essential alterations, encouraging the adoption of sustainable alternatives.

Author details

Prateek Singh and Jay Shankar Singh*
Department of Environmental Microbiology, Babasaheb Bhimrao Ambedkar University, Lucknow, Uttar Pradesh, India

*Address all correspondence to: jayshankar_1@yahoo.co.in

IntechOpen

References

[1] Tella TA, Festus B, Olaoluwa TD, Oladapo AS. Water and wastewater treatment in developed and developing countries: Present experience and future plans. In: Smart Nanomaterials for Environmental Applications. Netherlands: Elsevier; 2025. pp. 351-385

[2] Mumtaz M, Jahanzaib SH, Hussain W, Khan S, Youssef YM, Qaysi S, et al. Synergy of remote sensing and geospatial technologies to advance sustainable development goals for future coastal urbanization and environmental challenges in a riverine megacity. ISPRS International Journal of Geo-Information. 2025;**14**(1):30

[3] Seidu A, Salifu S, Sulemana I. Green Marketing Aspect of Green Supply Chain Management. Green Supply Chain Management; 2025

[4] Omemen KM, Aldbbah MO. Climate change: Key contributors and sustainable solutions. International Journal of Electrical Engineering and Sustainability. 2025;**3**(1):10-27

[5] Tripathi SN, Yadav S, Sharma K. Air pollution from biomass burning in India. Environmental Research Letters. 2024;**19**(7):073007

[6] Ogwu MC, Izah SC, Ntuli NR, Odubo TC. Food security complexities in the global south. In: Food Safety and Quality in the Global South. Springer Nature Singapore: Singapore; 2024. pp. 3-33

[7] Min H, Cho NJ. Waste management for environmentally sustainable cities: A quadruple helix collaboration in practice. ACS Sustainable Resource Management. 2024;**1**(8):1620-1629

[8] Sharma GK, Ghuge VV. How urban growth dynamics impact the air quality? A case of eight Indian metropolitan cities. Science of the Total Environment. 2024;**930**:172399

[9] Chen X, Yu L, Li Y, Liu T, Liu J, Peng D, et al. China's ongoing rural to urban transformation benefits the population but is not evenly spread. Communications Earth & Environment. 2024;**5**(1):416

[10] Rathod SV, Saras P, Gondaliya SM. Environmental pollution: Threats and challenges for management. In: Eco-Restoration of Polluted Environment. Boca Raton, Florida, USA: CRC Press; 2025. pp. 1-34

[11] Konrad T. Imagining Air: Cultural Axiology and the Politics of Invisibility. Exeter, United Kingdom: University of Exeter Press; 2023

[12] Atapattu AJ, Nuwarapaksha TD, Udumann SS, Dissanayaka NS. Integrated farming systems: A holistic approach to sustainable agriculture. In: Agricultural Diversification for Sustainable Food Production. Springer Nature Singapore: Singapore; 2025. pp. 89-127

[13] Sonet MS, Hasan MY, Kafy AA, Shobnom N. Spatiotemporal analysis of urban expansion, land use dynamics, and thermal characteristics in a rapidly growing megacity using remote sensing and machine learning techniques. Theoretical and Applied Climatology. 2025;**156**(2):79

[14] Nunes LJR. Reverse logistics as a catalyst for decarbonizing forest products supply chains. Logistics. 2025;**9**(1):17. DOI: 10.3390/logistics9010017

[15] Rani J, Gwal S. Challenges and opportunities in sustainable management of agrofood wastes. In: Sustainable Management of Agro-Food Waste. Oxford, United Kingdom: Elsevier; 2025. pp. 141-155

[16] Pratama RA. The correlation between air pollution and the prevalence of cardiovascular diseases in Jakarta. Studies in Social Science & Humanities. 2025;**4**(1):1-6

[17] Mallick J, Alqadhi S, Alkahtani M. Understanding the ecological health status of a semi-arid and arid region of Saudi Arabia in the era of rapid urbanization. Earth Systems and Environment. 2025:1-25

[18] Engdaw F, Fetahi T, Kifle D. Increasing anthropogenic stressors influenced the water quality and shifted trophic status of northern Lake Tana Gulf, Ethiopia. Heliyon. 2025;**11**(1). DOI: 10.1016/j.heliyon.2024.eXXXXX

[19] Ahmed AKA, El-Rawy M. The impact of aquifer recharge on groundwater quality. In: Managed Aquifer Recharge in MENA Countries: Developments, Applications, Challenges, Strategies, and Sustainability. Cham: Springer International Publishing; 2024. pp. 207-222. DOI: 10.1007/978-3-031-XXXXXX_13

[20] Saxena V. Water quality, air pollution, and climate change: Investigating the environmental impacts of industrialization and urbanization. Water, Air, & Soil Pollution. 2025;**236**(2):1-40. DOI: 10.1007/s11270-025-06543-0

[21] Yenalem T, Kidanie Y, Degu AM, Yisfa T. Investigating the effect of watershed management on land use, groundwater recharge, and irrigation potential in Tigray region, northern Ethiopia. International Journal of Hydrology Science and Technology. 2025;**19**(1):24-39

[22] Debnath S, Das Ghosh B, Lianthuamluaia L, Kumari S, Puthiyottil M, Karnatak G, et al. A hybrid ecological evaluation of the fisheries in changing climate: Case study from a peri-urban tropical wetland of Kolkata, Eastern India. Environmental Monitoring and Assessment. 2025;**197**(1):1-19. DOI: 10.1007/s10661-024-11897-2

[23] Elfaki MO, Kilic U. Determining quality properties, nutrient content, and relative feed value of tomato harvest waste silage. Placeholder. 2025

[24] Singh P, Verma A, Pratibha, Kumari A. A review on the environmental impact and management of sugar mill effluent through phytoremediation. Environmental Quality Management. 2025;**34**(3):e70027. DOI: 10.1002/tqem.70027

[25] Rawal S, Antany R, Kumar S, Pottakkal JG, Linda A. The perils of open landfill: A study on environmental risk assessment in Dharamshala, Himachal Pradesh, India. Environmental Monitoring and Assessment. 2025;**197**(2):177. DOI: 10.1007/s10661-024-11978-2

[26] Singh RK, Satyanarayana ANV. Aod Trends Over the Indo-Gangetic Plain During Last Two Decades: Impact of Land Use Changes and Crop Residue Burning. Available at SSRN 5081723

[27] Awos A. Environmental Assessment and Impact of Air, Water and Noise Levels near a Cement Factory in Ewekoro. Nigeria: Ogun State; 2024

[28] Najid N, Kouzbour S, Gourich B, Necibi MC, Elmidaoui A. Comparison of different membrane technologies

for Boron removal from seawater. In: Membrane Technologies for Heavy Metal Removal from Water. Boca Raton, Florida, USA: CRC Press; 2024. pp. 143-176

[29] Fanta B, Zemarku Z, Bojago E. Adoption of community-based land rehabilitation programs (CBLRP) and its effect on livelihoods in Offa district, South Ethiopia. Journal of Agriculture and Food Research. 2024;**16**:101104

[30] Kandel S. Advancing Resiliency Planning in Boston's Environmental Justice Communities: Introducing the Lived Experience of Dorchester's Caribbean Elders [Thesis]. Boston, Massachusetts, USA: University of Massachusetts Boston; 2023

[31] Ahmed M, Nasir AA, Masood M, Memon KA, Qureshi KK, Khan F, et al. Advancements in UAV-based integrated sensing and communication: A comprehensive survey. arXiv preprint arXiv:2501.06526. 2025

[32] Shamsuddoha M, Kashem MA, Nasir T. A review of transportation 5.0: Advancing sustainable mobility through intelligent technology and renewable energy. Future. Transportation. 2025;**5**(1):8

[33] Cai H. New York State Climate Impacts Assessment

[34] Juyal S, Naithani S, Gangopadhyay M. Air quality dynamics in North India. In: Climate Crisis and Sustainable Solutions: Strategies for Adaptation, Mitigation and Sustainable Development. Springer Nature Singapore: Singapore; 2024. pp. 195-209

[35] Upadhyay RK. Health hazards of various micro-pollutants, stubble smoke, furnace fumes and dust particles in urban areas. Global Research in Environment and Sustainability. 2023;**1**(1):57-81

[36] Ahn YJ, Koriyev M, Juraev Z. Urban soil dynamics: The relationship between soil health and urbanization. Journal of Asian Geography. 2024;**3**(2):62-69

[37] Duan H, Li Y, Yuan Y. A study on the long-term impact of crop rotation on soil health driven by big data. Geographical Research Bulletin. 2024;**3**:348-369

[38] Homeyer KL. Changing Tides: Integrating Nature-Based Solutions and Community Resilience in the Point & Collins Cove Neighborhoods [Thesis]. Medford, Massachusetts, USA: Tufts University; 2024

[39] Syracuse FG. Patterns of Government in Onondaga County: Structure and Services of County, City, Town, and Village Governments. FOCUS Greater Syracuse & Maxwell School, Syracuse University; 2024

[40] Hafiz M, Singh SJ, Pisini SK, Thammadi S. Islands at the Brink – Country Brief: Fiji (RECOVER Working Paper No. 2). CLARE Programme; 2024. Available from: https://clareprogramme.org/output/islands-at-the-brink-country-brief-fiji/

[41] Su Y, Zhang X, Xuan Y. Linking neighborhood green spaces to loneliness among elderly residents—A path analysis of social capital. Cities. 2024;**149**:104952

[42] Werner K. The need to (climate) adapts: Perceptions of German sports event planners on the imperative to address climate change. Frontiers in Sports and Active Living. 2024;**6**:1505372

[43] Akomolafe OO, Olorunsogo T, Anyanwu EC, Osasona F, Ogugua JO, Daraojimba OH. Air quality and public health: A review of urban pollution

sources and mitigation measures.
Engineering Science & Technology
Journal. 2024;**5**(2):259-271

[44] Pant H. Environment and
Agriculture: New Technological
Applications

[45] Edwards MR, Holloway T, Pierce RB,
Blank L, Broddle M, Choi E, et al.
Satellite data applications for sustainable
energy transitions. Frontiers in
Sustainability. 2022;**3**:910924

[46] Okoro PA, Chong K, Röder M.
Enabling modern bioenergy deployment
in Nigeria to support industry and local
communities. Biomass and Bioenergy.
2024;**190**:107403

An Agricultural Recovery Model for Rural Community Socioeconomic Development in Zimbabwe

Defe Rameck, Matutu Danmore and Mutote Karren

Abstract

The study sought to develop an agricultural recovery model for rural community socioeconomic development. A descriptive research design encompassing qualitative and quantitative data collection methods was adopted. Interviews, open-ended questionnaires, field observations, and secondary data sources were used to obtain qualitative data, and close-ended questionnaires were used to gather quantitative data. The research targeted households in Ward 32 of Zhombe in Kwekwe District and some stakeholders in the ward. As the selected study participants indicated, the findings revealed that agriculture contributes significantly to financial gains, food availability, employment opportunities, and community cooperation in the area. However, it was also established that some challenges hinder the existing agricultural practices to reach their maximum potential. The study concluded that agriculture contributes significantly to the livelihoods of households in agrarian-based communities, but it has not achieved its full potential due to identified challenges.

Keywords: agriculture recovery, livelihoods, rural communities, socioeconomic development, rural communities

1. Introduction

The agricultural sector is important in reviving the economic sector, as it addresses poverty and hunger globally [1]. In most developed countries, the contribution of agriculture to economic growth is occasionally overlooked or underrated and has long been seen as a minor economic sector [2]. This is because of the shift towards a service-based economy in developed countries and the rise in productivity of other sectors brought about by technological advances. While agriculture contributes to the economy in Germany, its contribution is not as significant as that of different industries [3]. Germany's gross domestic product will benefit 0.9% from agriculture in 2022 [4]. Various farming models, such as organic farming, which aims to increase yields, have been developed to improve productivity. Organic test farms generated profits in the 2020–2021 financial year, which were significantly greater than those of conventional farms [5].

The agricultural sector is the backbone and main driver of the most economically developed countries [6]. Developing nations largely depend on agriculture as the primary source of food, employment and income [7]. According to the United Nations Conference on Trade and Development (2022), 83% of the world's population lives in developing nations, with a significant portion of gross domestic product ranging from 30–60% generated from agricultural outputs, which also employs a considerable number of people, between 40 and 90% (FAO, 2022). As a developing country, India had agriculture contributing 16.17% of the country's gross domestic product (GDP) in 2015, 16.03% in 2018, 18.64% in 2020 and 16.62% in 2022 [4]. This decline in agricultural production was due to various factors, including COVID-19, which disrupted food systems, population growth, and the overexploitation of natural resources and poor cropping systems [8, 9].

In West Africa, for example, Nigeria, agriculture contributed 22.35% of the total GDP in 2021, with over 70% of the people in Nigeria engaging in the agriculture sector, particularly at a subsistence level [10]. In East Africa, for example, Kenya, over 40% of the workforce is employed in the agriculture industry, accounting for 70% of those living in rural areas and generating 33% of the country's GDP (World Bank, 2022). Agriculture accounted for 17.3% on average of the GDP of sub-Saharan Africa (SSA) in 2019 and 17.5% in 2022. This shows a slower rate of increase in agricultural productivity [10]. This is because most sub-Saharan Africa (SSA) countries' agriculture is rain-fed, and the region has recently experienced El Nino events, which have reduced agricultural production and hampered the socioeconomic development of SSA countries. Furthermore, the majority of countries in SSA are dominated by smallholder farmers who have inadequate irrigation, and 80% of agricultural output is produced by farmers who have an average of 2 hectares of land [11].

Agriculture is the backbone of the economy of Zimbabwe because the majority of Zimbabweans still live in rural areas and rely on agriculture and other related rural economic activities for their livelihood [12]. Small-scale farmers make up the majority of Zimbabwe's agricultural sector, with small-scale farmers taking up larger land areas situated in areas with infertile soils and unpredictable rainfall patterns (Runganga, 2021). The agricultural sector is essential for a country's development and economic growth, as shown by its 15–18% GDP contribution, 40% national export earnings, and 60% raw materials provided [13]. In 2019, agriculture contributed 9.8 and 8.8% to China's GDP in 2021, which indicates a decrease in agricultural production [4]. In recent years, various factors have contributed to the reduction of agricultural outputs, including climate change, land tenure and ownership issues, a lack of adequate government support, restricted access to technology, barriers to market access and economic instability [14–16]. The country's economy is mostly dependent on agriculture, which has reduced its GDP.

In Zimbabwe, a variety of agricultural models have been developed to increase production and improve the socioeconomic status of citizens. These models include models A1 and A2, which are part of the fast-track land reform programme [17]. Resettlement model A2 or commercialisation is currently being implemented in Zhombe North, and findings from [18] indicate that agricultural commercialisation affects household food security. However, Model A2 neglects other areas of Zhombe and focuses mostly on those groups of people who are assumed to be financially secure and a few chosen rural locations. This study aims to develop an agricultural recovery model for rural community socioeconomic development in Ward 32.

2. Materials and methods

2.1 Study area

This study was conducted in Ward 32 of Zhombe rural community (see **Figure 1**), which is located in the Kwekwe District in the Midlands Province [18]. Zhombe is in agroecological region III, which receives an average annual rainfall of 500–750 mm and an average temperature ranging between 18 and 24°C and has semi-intensive farming [19]. During the middle of the season, the region experiences intense dry spells, which are crucial for the growth of crops such as cotton and maize. The areas are dominated by paraferrallitic soils and coarse-grained sandy soils from granite [12]. The soils in the area are moderate to strongly leached soils with significant levels of free sesquioxide of iron and aluminium. The soils are resistant to erosion and have the potential to store significant amounts of water, which indicates that the soils are suitable for agricultural activities. Donjane has a total population of 6589, which heavily relies on subsistence agriculture (groundnuts, maize, round nuts, sorghum, millet, and sunflower) and livestock production (cattle, goats, and poultry).

2.2 Sampling design and data collection techniques

The target population consisted of 6589 people in Ward 32 because they accurately represent community interests. This study also targeted the District Development Coordinator, the Ward councillor and the Chief because they have resided in the area

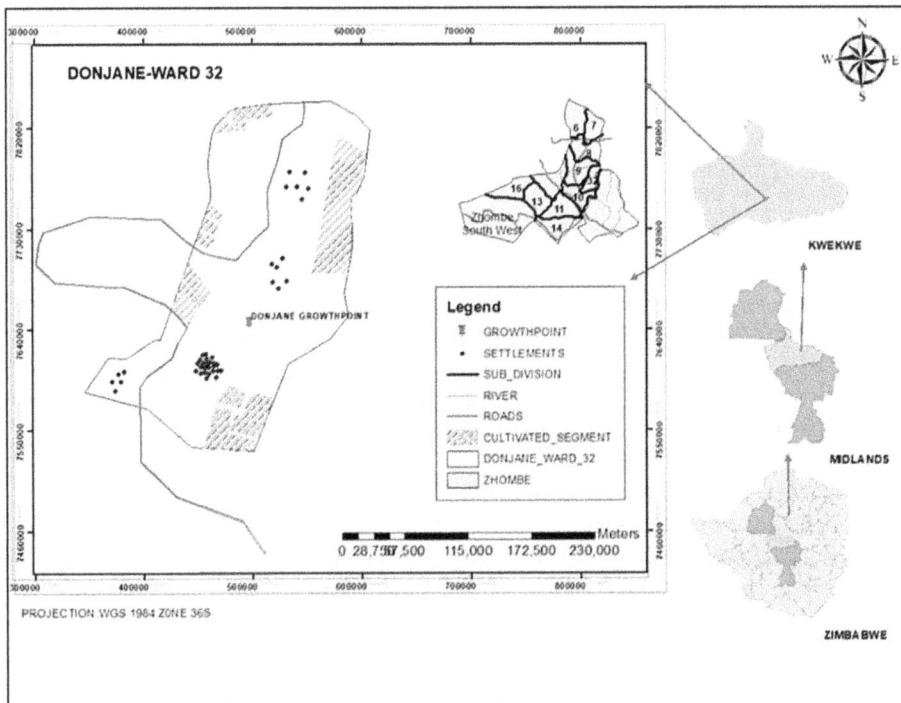

Figure 1.
Map of the ward 32 study area. Source: Authors 2024.

for a long time; thus, they have vast knowledge of the study area. The researcher also targeted the Agricultural Extension Officer to gain insights and expertise in sustainable agricultural practices. Livestock and veterinary officers were interviewed to gain more information on the diseases that could endanger the lives of livestock and the right vaccines to be used. As a result, this successfully achieved the objective of developing an agricultural recovery model for rural community socioeconomic development.

The sample size was determined *via* the Yamane1967 formula to increase the degree of accuracy when determining the fraction of a population to be sampled within a suitable margin of error [20].

$$n = \frac{N}{1 + N(e)^2} \qquad (1)$$

where n = the sample size.
N = Population size.
e = desired margin of error.
n = 6589/1 + 6589(0.12)2.
n = 68 sample size.

The study used a systematic sampling technique to select 68 participants for questionnaires. Systematic sampling is a statistical sampling technique using predetermined intervals to sample a large population [21]. This method ensures that the sample contains every ninth person in the total population, producing a representative subset of the entire population [22]. The study also used a purposive sampling method to identify five key informants for interviews, including the Agritex officer, the ward councillor, the Traditional Chief, the Livestock and Veterinary Officer, and the District Development Committee (DDC). An agricultural recovery model for rural community socioeconomic development was developed based on the responses of the five key informants. The quantitative data obtained was subjected to descriptive analysis using SPSS version 22.0, while qualitative data was subjected to content analysis.

3. Results and discussion

3.1 Existing agricultural practices for socioeconomic development in Zhombe

Agricultural practices such as zero tillage, mulching, mixed farming, monoculture, and cattle farming are being implemented in the study area. Local farmers cultivate various crops, including maize, ground nuts, round nuts, sorghum, millet, sunflower, sugar beans, and cowpeas. A majority of farmers in Ward 32 cultivate drought-resistant crops because of their adaptability to rainfall patterns and local soils. Participants mentioned that they could sell almost half of their produce, mostly in the Central Business District (CBD) market of Kwekwe, and use the income to increase their standards of living and purchase the necessary supplies needed for the upcoming growing seasons. It was revealed during an interview with the Agritex officer that local farmers also obtain economic benefits from selling their livestock and poultry. Melo et al. [23] emphasised that through practising agriculture, farmers can achieve socioeconomic benefits in the form of financial gains that will help their livelihoods.

Households primarily cultivate crops for consumption and sell excess crops only when their food needs are met. The interviewed Ward councillor highlighted that agricultural practices contribute to food availability in the community, especially when crops cultivated are mainly for subsistence rather than as cash crops. Local people heavily depend on their agricultural produce to feed their families, and these methods are crucial in fighting poverty. According to Simbarashe et al. [24], a large part of agricultural production in rural communities is mainly for family consumption. The interviewed veterinary officer stated that most families in the ward keep livestock mainly for selling more than for consumption.

Of the total respondents selected for this study, 17% highlighted that agriculture is the main source of employment and alleviates the area's high rate of unemployment and shortage of formal job opportunities. Respondents emphasised that agricultural activities have created job opportunities for crop growers, farm workers, livestock caretakers, and poultry handlers. The diverse agricultural methods in Ward 32 offer a variety of job opportunities, thus enabling residents to contribute to the farming industry and sustain their livelihoods. Agricultural activities have a positive impact on the local economy which generates more employment opportunities. In an interview with the DDC official for Ward 32, the marketing sector, transportation, and agro-processing industries were mentioned as areas where agricultural activity has immensely created job possibilities. According to Nyamapfene [12], agriculture provides a stable source of income for the community since there is a shortage of formal job opportunities.

3.2 Challenges of existing agricultural practices

A total of 43% of households mentioned sporadic rainfall patterns as a major contributing factor to poor yields. Erratic rainfall patterns have accelerated the frequency of dry seasons. During an interview, the Agritex officer emphasised the importance of rainfall in improving Zhombe Ward 32's agricultural practices highlighting its adverse impacts on livelihoods and livestock. Agriculture activities are greatly affected by erratic and unpredictable rainfall patterns caused by climate change and other factors. Local farmers highlighted that uncertain rainfall patterns make it difficult to effectively plan and carry out agricultural activities. Control measures for crop growth and water availability are all affected by this uncertainty [25].

Respondents mentioned that the scarcity of resources such as raw materials, expertise, knowledge, water infrastructure, and financial capacity to buy essential medicine for their livestock are the main challenges in Zhombe. Respondents highlighted that securing sufficient raw materials such as seeds, fertiliser, and farming equipment required for their agricultural practices is challenging. Furthermore, the lack of access to knowledge and training on current sustainable agriculture practices makes it more difficult to utilise more advanced farming methods, reducing productivity and increasing the risk of crop failure.

Inadequate water infrastructure and water sources in Ward 32 have greatly affected its socioeconomic development. Respondents highlighted that one dam in the area has been dry for a long time, and two boreholes are situated far from where the residents live, making it more difficult for them to rely upon available water sources for agricultural practices.

Approximately 9% of the participants indicated that the lack of arable land significantly constrains agricultural yields and productivity. Participants raised concerns about decreasing availability of land appropriate for cultivation, making it more difficult for them to produce sufficient food crops to support their livelihoods. Food scarcity is attributed to a limited amount of arable land, soil degradation, and pressure from the growing population. The Agritex officer highlighted that there is not enough land to support the ward population. The lack of fertile land in the area makes it essential to put policies in place to maximise land use and enhance agricultural productivity while using available resources. These results highlight the importance of comprehensive strategies that can address challenges affecting land availability, productivity, and sustainable methods for managing land to improve agricultural outputs and help in the community's socioeconomic development [1, 26].

The quantitative information gathered from the respondents indicates that 10% of the community's households lack the required support systems to improve agricultural mechanisms. The participants mentioned that their community's remote location makes it challenging to access government support and services for mechanisation. The Agritex officer explained that residents in Zhombe encounter challenges in promptly receiving support, direction, and materials because of their remote location. Accessing services such as veterinary treatment, agronomic guidance, and extension services requires travelling long distances on inadequate roads, which poses a significant challenge for farmers. The Agritex officer emphasised the need to improve access to veterinary services, agronomists, and other agricultural experts who can guide crop management, pest control, and livestock healthcare. For the past years, the absence of these essential services has hampered the ability of the community to address agricultural challenges effectively and maximise productivity.

3.3 Perceptions between agriculture and community socioeconomic development

Observations indicate that participants involved in crop diversification positively perceived crop diversification as a part of the socioeconomic development of their communities. Concerning crop diversity, 47% of households strongly agree that it is favourable for socioeconomic development since it improves food security and provides a more consistent income. Crop diversification results in the availability of a variety of agricultural products, which improves household income and enhances household livelihoods. The Agritex officer strongly agreed that crop diversification boosts household income and improves food security as it minimises the effects of diseases, pests, and climate variability. During interviews, participants strongly agreed that crop diversification helps them have a variety of food options, which contributes to their balanced diet. Cultivating multiple crops enhances access to a variety of markets, expands income sources, and improves the community's overall resilience. Approximately 10% of the respondents strongly disagreed that crop diversification has improved food security and household income in the ward these past years because they perceive monocultures as more effective since they produce higher yields.

Results indicate that 31% of the respondents agreed that organic farming practices such as composting and organic fertilisers improve agricultural productivity to enhance food security. The participants mentioned the importance of minimising chemical fertilisers and their negative effects on human health and the environment. Participants perceive organic farming as a sustainable method that improves the health of the soil and produces healthier crops and the use of manure and cow dung is an efficient substitute for chemical fertilisers. Additionally, participants agreed that ash and soil mould are efficient substitutes to improve crop production and soil moisture to enhance productivity. The use of organic farming practices effectively contributes to the socioeconomic development of the community; however, participants positively believe that the use of chemical fertilisers such as Compound D can increase crop production and yields. Compound D is a type of fertiliser that releases nutrients continuously over time, which is a characteristic of controlled-release fertilisers (CRFs) designed to provide a steady supply of nutrients to plants, improving nutrient use efficiency and reducing environmental impact.

Over 47% of the participants strongly agreed that irrigation systems and water harvesting techniques can improve and address food security issues in the community. Interviewed councillors highlighted that in households with access to reliable water supply and adopted effective irrigation techniques there is potential to increase income and food production.

During key informant interviews, respondents highlighted that increasing production, minimising labour requirements, and improving the quality of crops can be achieved through the use of mechanisation and improved seeds. Over 40% of the households in Zhombe strongly believe that modern technologies can increase agricultural productivity and profitability; however, several participants are still using traditional farming methods such as zero tillage and cattle farming, and in their view, these methods are cost-effective, require fewer inputs, and produce more yields.

3.4 An agricultural recovery model for socioeconomic development

Based on the research findings, the study synthesised an agricultural recovery model for rural socioeconomic development (see **Figure 2**). Key informants such as representatives from Agritex, the Livestock and Veterinary Officer, the Chief, the Councillor, and the DDC provided information that helped create the model. An agricultural recovery model is a structured approach or framework that seeks to revive and improve the agricultural systems and practices in areas or communities that have encountered disturbances or obstacles including natural disasters, political unrest, or economic crises [10].

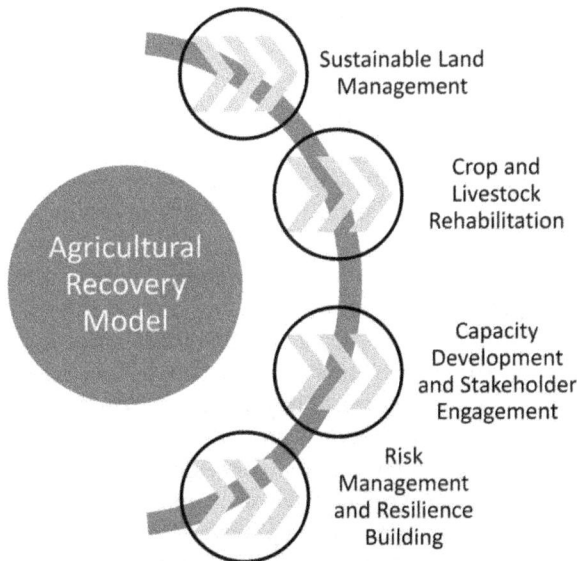

Figure 2.
Agricultural recovery model.

Crop and livestock rehabilitation are key components in assisting farmers to recover their crop and livestock production. The process includes repairing or reconstructing damaged agricultural infrastructure, including irrigation systems, farm buildings, storage facilities, roads, and marketplaces. Rehabilitating infrastructure is crucial for restoring production capacity, improving market access, and enhancing overall agricultural productivity. Other activities may include providing improved and disease-resistant seeds, offering training on modern farming techniques, promoting sustainable land management practices, and supporting livestock restocking programmes. Facilitating market linkages and creating opportunities for farmers to sell their products are very important. This can involve establishing farmer cooperatives, promoting value-added processing, improving postharvest handling and storage practices, and facilitating access to local and regional markets.

3.4.1 Risk management and resilience building

Risk management and resilience building are instrumental in enhancing the resilience of the agricultural sector to future shocks and stresses. Implementing agricultural risk management strategies in Zimbabwe requires a multi-faceted approach that addresses the unique challenges faced by farmers in the region. There is a need to enhance access to diversified crop and livestock varieties that are resilient to climate change. Activities may include promoting climate-smart agriculture, implementing disaster risk reduction strategies, providing insurance schemes, and facilitating access to credit and financial services for farmers. For the past few years, smart agriculture has played a transformative role for large-scale farmers in Zimbabwe by enhancing productivity and resilience. With the use of technology such as GPS, IoT sensors, and mobile applications, farmers can monitor soil health, optimise water usage, and make data-driven decisions that lead to increased yields and reduced costs. To ensure the sustainability of the established mechanisms, the progress and impact of the agricultural recovery model should be monitored. This involves collecting data, measuring key performance indicators, conducting impact assessments, and using the findings to make informed adjustments and improvements to recovery strategies.

3.4.2 Capacity development and stakeholder engagement

Capacity development and stakeholder engagement are critical in building the technical and managerial capacities of farmers, extension workers, and other stakeholders involved in the recovery process. An effective agricultural framework aims at enhancing productivity and sustainability by equipping farmers, extension workers, and local institutions with the skills, knowledge, and resources necessary to adopt innovative practices and technologies. Collaborating with relevant stakeholders such as farmers, local communities, government agencies, nongovernmental organisations (NGOs) and private sector entities is key to building capacity. Engaging these stakeholders ensures that their perspectives, knowledge, and needs are considered in the recovery planning process. This inclusive strategy enhances the relevance and effectiveness of interventions and builds trust and ownership within communities, leading to more sustainable agricultural practices and improved food security. Combining capacity development with stakeholder engagement creates a resilient agricultural system that can adapt to changing conditions and meet the needs of diverse populations.

3.4.3 Crop and livestock rehabilitation

Crop and livestock rehabilitation is essential for restoring agricultural productivity. This comprehensive approach instigates a thorough assessment of damage to crops and livestock, engaging local farmers and stakeholders to gather insights. Based on these evaluations, tailored rehabilitation strategies are developed, focusing on soil restoration, the introduction of resilient crop varieties, and enhanced livestock care. Effective water management practices, including improved irrigation and rainwater harvesting, are also implemented to ensure adequate resources. Throughout the process, continuous monitoring and community engagement ensure that the rehabilitation efforts are effective and aligned with local needs, ultimately building resilience and enhancing food security for the agricultural sector.

3.4.4 Sustainable land management

Sustainable land management practices help conserve and improve the quality of soil. By minimising soil erosion, maintaining soil fertility and preventing degradation, these practices ensure the long-term productivity of agricultural land. This is particularly important in the context of agricultural recovery, as degraded or eroded soil can significantly hamper the ability to restore productivity and meet food demand. Sustainable land management practices are essential in an agricultural recovery model because they ensure the long-term productivity, resilience, and economic viability of farming systems. By focusing on soil water management, climate change mitigation, and economic stability, sustainable land management practices contribute to a sustainable and resilient agricultural sector.

4. Conclusion

The main goal of this research was to develop an agricultural recovery model for rural community socioeconomic development in Ward 32, Zhombe. The results indicated that agriculture has a positive impact on the livelihoods and well-being of the residents in Ward 32. The data show that agricultural production contributes to food security, household income, employment, and social cohesion. However, owing to the challenges they face, it is clear that their agricultural production is declining. These challenges have prevented households from maximising their agricultural production. The study also revealed that most households strongly agree that agricultural production contributes to socioeconomic development in the area.

Author details

Defe Rameck[1], Matutu Danmore[2*] and Mutote Karren[1]

1 Department of Geography, Environmental Sustainability and Resilience, Midlands State University, Gweru, Zimbabwe

2 Department of Community and Social Development, University of Zimbabwe, Harare, Zimbabwe

*Address all correspondence to: danmore46@gmail.com

IntechOpen

References

[1] Pawlak K, Kołodziejczak M. The role of agriculture in ensuring food security in developing countries: Considerations in the context of the problem of sustainable food production. Sustainability. 2020;**12**(13):5488. DOI: 10.3390/su12135488

[2] Loizou E, Karelakis C, Galanopoulos K, Mattas K. The role of agriculture as a development tool for a regional economy. Agricultural Systems. 2019;**173**:482-490. DOI: 10.1016/j.agsy.2019.04.002

[3] Mohr S, Kühl R. Acceptance of artificial intelligence in German agriculture: An application of the technology acceptance model and the theory of planned behavior. Precision Agriculture. 2021;**22**(6):1816-1844. DOI: 10.1007/s11119-021-09814-x

[4] World Bank, Agriculture, Forestry, and Fishing, Value Added (% of GDP), World Bank National Accounts Data, and OECD National Accounts Data Files. Available from: https://data.worldbank.org/indicator/nv.agr.totl.zs?most_recent_year_desc=false

[5] Federal Ministry of Food and Agriculture. Organic Farming in Germany, Food and Agriculture Organization (2020) Zimbabwe at a Glance. Germany: Federal Ministry of Food and Agriculture; 2022. pp. 12-15. Available from: https://www.fao.org/zimbabwe/fao-in-zimbabwe/zimbabwe-at-a-glance/en/

[6] Girma Y, Kuma B. A meta-analysis on the effect of agricultural extension on farmers' market participation in Ethiopia. Journal of Agriculture and Food Research. 2022;**7**:100253. DOI: 10.1016/j.jafr.2021.100253

[7] Hemathilake DMKS, Gunathilake DMCC. Chapter 31—Agricultural Productivity and Food Supply to Meet Increased Demands. In: Bhat R., editor. Future Foods: Global Trends, Opportunities, and Sustainability Challenges. London: Academic Press; 2022:539-553. DOI: 10.1016/B978-0-323-91001-9.00016-5

[8] Harris J et al. Food system disruption: Initial livelihood and dietary effects of COVID-19 on vegetable producers in India. Food Security. 2020;**12**:841-851. DOI: 10.1007/s12571-020-01064-5

[9] Sathish P. Rural non-farm income in the rural areas of Punjab. Access and Determinants. 2022;**79**(3):18

[10] Food and Agriculture Organization. FAOSTAT Statistical Database of the United Food and Agriculture Organization. Rome: Food and Agriculture Organization; 2022

[11] Jellason NP, Robinson EJZ, Ogbaga CC. Agriculture 4.0: Is sub-Saharan Africa ready? Applied Sciences. 2021;**11**(12):5750. DOI: 10.3390/app11125750

[12] Nyamapfene K. A geographical overview of the soils of Zimbabwe and their agricultural potential. Geographical Association of Zimbabwe. 1992;**15**(1):63-73

[13] Mujeyi A, Mudhara M, Mutenje M. The impact of climate smart agriculture on household welfare in smallholder integrated crop–livestock farming systems: Evidence from Zimbabwe. Agriculture & Food Security. 2021;**10**(1):4. DOI: 10.1186/s40066-020-00277-3

[14] Frischen J, Meza I, Rupp D, Wietler K, Hagenlocher M. Drought risk to agricultural systems in Zimbabwe: A spatial analysis of Hazard. Exposure, and Vulnerability, Sustainability. 2020;**12**(3):752. DOI: 10.3390/su12030752

[15] Nhemachena C, Nhamo L, Matchaya G, Nhemachena CR, Muchara B, Karuaihe ST, et al. Climate Change Impacts on Water and Agriculture Sectors in Southern Africa: Threats and Opportunities for Sustainable Development. Water. 2020;**12**(10):2673. DOI: 10.3390/w12102673

[16] Scoones I, Mavedzenge B, Murimbarimba F. Young people and land in Zimbabwe: Livelihood challenges after land reform. Review of African Political Economy. 2019;**46**(159):117-134. DOI: 10.1080/03056244.2019.1610938

[17] Nyawo VZ. The fast track land reform of Zimbabwe read through the lens of Ubuntu. Third World Thematics: A TWQ Journal. 2023;**8**(4-6):189-204. DOI: 10.1080/23802014.2023.2196980

[18] Madududu P, Ndayitwayeko W-M, Mwakiwa E, Mutambara J. Impact of agricultural commercialisation on household food security among smallholder farmers in Zhombe north Rural District, Zimbabwe. East African Journal of Science, Technology and Innovation. East African Science and Technology Commission - EASTECO KG; 2021;**2**. DOI: 10.37425/eajsti.v2i2.244

[19] Vincent V, Thomas RG. An Agricultural Survey of Southern Rhodesia Part 1: Agro-Ecological Survey. Salisbury: Government Printers; 1960

[20] Adam AM. Sample size determination in survey research. Journal of Scientific Research and Reports. 2020;**26**:90-97. DOI: 10.9734/jsrr/2020/v26i530263

[21] Akpan B, Ebenezer E, Collins Piate R. Assessment of different methods of sampling techniques: The strengths and weakness. Shared Seasoned International Journal of Topical. 2023;**9**(1):64-83

[22] Makwana D, Engineer P, Dabhi A, Chudasama H. Sampling methods in research: A review. International Journal of Trend in Scientific Research and Development. 2023;**7**:762-768

[23] Melo FPL et al. Adding forests to the water–energy–food nexus. Nature Sustainability. 2021;**4**(2):85-92. DOI: 10.1038/s41893-020-00608-z

[24] Simbarashe M, Runganga R, Mhaka S. Impact of Agricultural Production on Economic Growth in Zimbabwe. Munich Personal RePEc Archive. Germany: University Library of Munich; 2021

[25] Chivasa N. Sustainability of food production support services offered by sustainable agriculture trust to subsistence farmers in Bikita District, Zimbabwe. Jamba: Journal of Disaster Risk Studies. 2019;**11**(1):1-9. DOI: 10.4102/jamba.v11i1.526

[26] Muchesa E. Perceptions and experiences regarding the current market system by communal farmers in Mhondoro-Mubaira (Zimbabwe). International Journal of Sustainable Development Research. 2018;**4**(2):31. DOI: 10.11648/j.ijsdr.20180402.13

www.ingramcontent.com/pod-product-compliance
Lightning Source LLC
Chambersburg PA
CBHW081335190326
41458CB00018B/6012